Forest and Plant Bioproducts Division

Core Programming Topic at the 2012 AIChE Annual Meeting

Pittsburgh, Pennsylvania, USA
28 October – 2 November 2012

ISBN: 978-1-62276-729-8

Printed from e-media with permission by:

Curran Associates, Inc.
57 Morehouse Lane
Red Hook, NY 12571

Some format issues inherent in the e-media version may also appear in this print version.

Printed by Curran Associates, Inc. (2013)

For permission requests, please contact AIChE
at the address below.

AIChE
3 Park Avenue
New York, NY 10016-5991

Phone: (203) 702-7660
Fax: (203) 775-5177

www.aiche.org

Additional copies of this publication are available from:

Curran Associates, Inc.
57 Morehouse Lane
Red Hook, NY 12571 USA
Phone: 845-758-0400
Fax: 845-758-2634
Email: curran@proceedings.com
Web: www.proceedings.com

TABLE OF CONTENTS

(219a) "Science Looks to Unlock Potential In the Undomesticated Tree Through Precise Genetic Breeding"

Lawrence Sullivan, Bioenergy, ArborGen, Inc., Ridgeville, SC

Scientific advances are unlocking ever more significant uses for the components of trees. In addition to the wood, pulp for paper and wood for fire, increasing demands are being made of trees for fine grade cellulosic sugars that can be converted to biofuels, polyethylene and other compounds for industrial use. Yet the tree itself remains an undomesticated plant. Far from its agricultural counterparts which have been bred for cold tolerance, drought resistance, improved yields and improved food quality, the tree has only been bred for a few generations and predominately to improve form for processing lumber.

ArborGen uses advance breeding techniques to shorten the time required to reach domesticated species for industrial purposes. Varietal propagation allows us to replicate one species to improve consistency. Mass Controlled Pollination allows us to improve a line by controlling one parent in a breeding pair. However, demands for wood cellulose are growing at a rate of acceleration that can never be matched at the conventional breeding level for a species that takes more than 20 years to mature. ArborGen scientists use precision breeding techniques to look at specific ways of improving and domesticating trees for industrial uses.

One excellent example is a lignin-modified *Eucalyptus* species that has proven to release more than twice the usual amount of sugar, making it a promising option as a biomass feedstock for liquid fuel. The tree is being studied by National Renewable Energy Lab (NREL). Using plant biotechnology the modifications were made at two points in the lignin biosynthetic pathway, with the largest increase in sugar release coming from cinnamate-4-hydroxylase (C4H) down-regulation. Although some "low recalcitrance" plant lines suffer from reduced growth, many of the C4H down-regulated lines from the *E. grandis x E. urophylla* cross grow well. C4H lines have an estimated biomass productivity of ten dry-tons per acre per year, with the potential to produce about 1,000 gallons of liquid biofuels per acre.

Scientists from NREL have characterized the C4H lines as containing half the lignin of the unmodified lines. Using a high throughput sugar release assay developed at NREL (Selig et al, 2011 *Biotech. Letters* 1-7), the modified lines were found to release up to 99 percent of their sugars, up from 40 percent in the non-modified plants. This result translates to an improvement of 150 percent, a dramatic demonstration of the impact of lowering recalcitrance.

Further work by NREL and ArborGen on these and similar lines will aim to understand exactly how the recalcitrance is lowered and how this knowledge can be used to develop healthy low recalcitrance lines in an array of species.

(219b) Process Synthesis and Global Optimization of Biomass to Liquid Transportation Fuels (BTL) Systems

Richard C. Baliban, Josephine A. Elia and Christodoulos A. Floudas, Chemical and Biological Engineering, Princeton University, Princeton, NJ

The transportation sector relies heavily on petroleum as a primary energy source, and currently faces challenges over high crude oil prices, volatility of the global oil market, and high levels of greenhouse gas (GHG) emissions [1]. As a result, a substantial effort has been taken in the United States to focus on greater energy independence through the utilization of more domestic energy sources. A recent review highlights the single and hybrid feedstock design alternatives that can produce gasoline, diesel, and kerosene from coal, biomass, or natural gas [2]. Of particular interest are liquid fuels that can be derived from biomass, since the feedstock has the potential to both a renewable energy source and a large contributor to a reduction in the well-to-wheel GHG emissions of the transportation sector. Though corn-based ethanol and soybean-based diesel comprise a majority of the biofuels manufactured today, their use for fuel production has led to concerns regarding the impact on the price and availability of these feedstocks as sources of food [3]. Lignocellulosic plant sources including dedicated energy crops, agricultural crop residues, and forest residues are capable of being sustainable harvested without affecting the food chain and therefore expected to be a more considerable source of biofuels in the future. The development of processes that can economically utilize these feedstocks to produce synthetic liquid fuels will be of significant importance.

We present an optimization based framework to perform a comprehensive technoeconomic, environmental, and societal assessment of an alternative energy refinery that will convert biomass to the liquid transportation fuels (BTL) gasoline, diesel, and kerosene. Rigorous process design, global optimization, and process synthesis strategies will be utilized [4-11] to determine the optimal plant topology of a BTL refinery that minimizes the overall cost of liquid fuels production. The cost of utility generation and wastewater treatment is treated using a simultaneous heat, power, and water integration within the process synthesis model. Multiple case studies are presented to investigate the effect of biomass type, biomass moisture content, plant capacity, and product distribution on the optimal process topology, total plant cost, and break-even oil price. Major topological decisions from the optimization model and the result on plant capital cost, total operational cost, and the process utility network will be discussed.

[1] National Academy of Sciences, National Academy of Engineering, and National Research Council. Liquid Transportation Fuels from Coal and Biomass: Technological Status, Costs, and Environmental Issues. Washington, D. C., EPA, 2009.

[2] C.A. Floudas, J.A. Elia, R.C Baliban. Hybrid and Single Feedstock Energy Processes for Liquid Transportation Fuels: A Critical Review. *Comp. Chem. Eng.*, 2012: 41(11), 24-51.

[3] L.R. Lynd, E. Larson, N. Greene, M. Laser, J. Sheehan, B.E. Dale, S. McLangulin, M. Wang. The role of biomass in America's energy future: framing the analysis. *Biofuels, Bioprod. Biorefin.*, 2009:3(2), 113.

[4] R.C. Baliban, J.A. Elia, C.A. Floudas. Toward Novel Hybrid Biomass, Coal, and Natural Gas Processes for Satisfying Current Transportation Fuel Demands, 1: Process Alternatives, Gasification Modeling, Process Simulation, and Economic Analysis. *Ind. Eng. Chem. Res.,* 2010:49(16), 7343-7370.

[5] J.A. Elia, R.C. Baliban, C.A. Floudas. Toward Novel Hybrid Biomass, Coal, and Natural Gas Processes for Satisfying Current Transportation Fuel Demands, 2: Simultaneous Heat and Power Integration. *Ind. Eng. Chem. Res.*, 2010:49(16), 7371-7388.

[6] R.C. Baliban, J.A. Elia, C.A. Floudas. Optimization Framework for the Simultaneous Process Synthesis, Heat and Power Integration of a Thermochemical Hybrid Biomass, Coal, and Natural Gas Facility. *Comp. Chem. Eng.*, 2011:35(9), 1647-1690.

[7] J. A. Elia, R.C. Baliban, X. Xiao, C.A. Floudas. Optimal Energy Supply Network Determination and Life Cycle Analysis for Hybrid Coal, Biomass, and Natural Gas to Liquid (CBGTL) Plants Using Carbon-based Hydrogen Production. *Comp. Chem. Eng.*, 2011:35(8), 1399-1430.

[8] R.C. Baliban, J.A. Elia, C.A. Floudas. Simultaneous Process Synthesis, Heat, Power, and Water Integration of Thermochemical Hybrid Biomass, Coal, and Natural Gas Facilities. *Comp. Chem. Eng.*, 2012:37(10), 297.

[9] R.C. Baliban, J.A. Elia, R. Misener, C.A. Floudas. Global Optimization of a Thermochemical Based Hybrial Coal, Biomass, and Natural Gas to Liquids Facility. *Comp. Chem. Eng.*, 2012, http://dx.doi.org/10.1016/j.compchemeng.2012.03.008.

[10] R.C. Baliban, J.A. Elia, V.W. Weekman, C.A. Floudas. Process synthesis of hybrid coal, biomass, and natural gas to liquids via Fischer-Tropsch synthesis, ZSM-5 catalytic conversion, methanol synthesis, methanol-to-gasoline, and methanol-to-olefins/distillate technologies. *Comp. Chem. Eng.*, submitted.

[11] J.A. Eila, R.C. Baliban, C.A. Floudas. Nationwide Energy Supply Chain Analysis for Hybrid Feedstock Processes with Significant CO_2 Emissions Reduction. *AIChE J.*, submitted.

CONDUCTIVE LAYERS ON SURFACE MODIFIED NATURAL FIBRE BASED SUBSTRATES FOR PRINTED FUNCTIONALITY

Dimitar Valtakari[1], Roger Bollström[1], Mikko Tuominen[2], Hannu Teisala[2], Mikko Aromaa[3], Martti Toivakka[1], Jurkka Kuusipalo[2], Jyrki M. Mäkelä[3], Jun Uozumi[4], and Jarkko J. Saarinen[1,4,]*

[1] Laboratory of Paper Coating and Converting, Center for Functional Materials, Abo Akademi University, Turku, FINLAND
[2] Paper Converting and Packaging Technology, Tampere University of Technology, Tampere, FINLAND
[3] Aerosol Physics Laboratory, Tampere University of Technology, Tampere, FINLAND
[4] Faculty of Engineering, Hokkai-Gakuen University, Sapporo, JAPAN
[*] Corresponding author: jarkko.j.saarinen@abo.fi

Abstract

Formation of conductive surfaces by flexographical printing has been studied using an IGT test printer with PEDOT-PSS and Ag conductive inks on coated papers. Printability of multilayer coated paper and TiO_2 nanoparticle coating generated by the liquid flame spray process are compared to commercial plastic film used in printed electronics applications. The wettability of TiO_2 nanoparticle coating can be altered between superhydrophobic and superhydrophilic states by ultraviolet light. It is observed that superhydrophobicity induced by TiO_2 nanoparticles results in poorer ink setting and hence lower conductivities with water-based PEDOT:PSS ink. Therefore, we observe conductivity only after several successive prints. On contrary, we observe several orders of magnitude better conductivities when using a silver ink in flexography. It is believed that sustainable natural fibre based substrates will find more applications in printed electronics application in the future.

Introduction

Printed electronics and intelligence have been under a growing interest since the start of the 21st century with the market value forecasts up to 300 billion USD by 2025 [1]. Conventionally such applications have been produced on plastic films. However, paper based electronics [2] has been studied recently, and it has been shown that simple, all-printed transistors can be made on multilayer coated paper [3]. Another example is a thin, lightweight, and foldable thermochromic display based on a regular copy paper in which thermochromic ink was utilized with patterned electrodes on the other side of the paper providing extremely low cost displays with price less than 0.10 USD/m^2 [4]. Such displays may find applications, for example, in data reporting on portable point-of-care devices. Natural fibre based substrates are rapidly finding new applications outside of the conventional graphical arts industry.

Paper has many advantages over the plastic films: it is made of renewable materials and it is cheap with tailorable surface properties. Paper can be 1000 times cheaper than glass substrates and 100 times cheaper than plastic films [2]. Cellulose, the major component of plant biomass, is the most abundant biopolymer on Earth with annual production up to 1.8×10^{12} tonnes [5]. However, compared to plastic films, paper is porous, uneven, and rough network of fibres. Typically paper surface is smoothened by dispersion coating consisting mineral pigments and organic binders. Recently, nanoscale coating techniques including layer-by-layer and liquid flame spray coating have been studied that allow surface functionalization with significantly reduced coating amounts. Nanoscale control of surface properties is crucial for achieving good performance in electronics applications.

This study concentrates on printing large area conductive surfaces on various coated paper grades. Such conductive surfaces are needed, for example, in electrochromic displays and in solar cells. Photovoltaic (PV) devices on paper substrates were demonstrated by two groups in 2011: Karen Gleason's group [6] utilized an oxidative chemical vapour deposition (oCVD) for fabrication of PV cells on rough tissue paper. The key for the success of such an approach was perfect contour coating by oCVD preventing formation of electrical shorts. However, the process is a slow batch process and not optimal as paper is produced in practice with line speeds typically above 1000 m/min.

An alternative approach was presented by Arvid Hübler's group [7] based on all-printed process flow utilizing a transfer of bottom electrode from a plastic film. With such an approach they were able to overcome the roughness issues of paper as the plastic film defines the roughness. However, their approach, although based on all-printing processes, did not allow a continuous process flow but rather had up to 12 hour gaps between different prints. The observed power conversion efficiency was 1.31 %. For cost-effective manufacturing of organic electronic devices, it would be beneficial to use an all-in-line printing process with possibilities for significant unit cost reduction of the devices [8].

In current PV applications indium tin oxide (ITO) is typically used as a transparent conductive electrode. Unfortunately, there are two major drawbacks with ITO: first, it is expensive and scarce material. Secondly, the brittle nature of the ITO films makes them not suitable for flexible substrates. These limitations prevent the use of ITO on low-cost, large-scale PV applications. Conductive polymers such as poly(3,4-ethylenedioxythiophene)/poly(styrenesulfonate) (PEDOT/PSS) may be suitable candidates as they can be both transparent and conductive. However, there is a strong correlation between conductivity and opacity *i.e.* the better the conductivity, the more opaque is the film. This has been a bottleneck but recent developments have eased the problem of achieving good transparency with high conductivity [9].

PEDOT:PSS is the most common and affordable conductive polymer currently available on the market. It comes in 1:2.5 (pH 1.5 – 2.5), 1:6 (pH 4.0 – 7.0) or 1:20 (pH 7.0) ratio by weight with a solids content of 0.6 – 3.4 % dispersed in water. Viscosity can be varied from 4 to 350 mPas depending on requirements of the application. PEDOT:PSS typically serves as a transparent conductive electrode or as a buffer layer in between the active layer and metallic electrode. Conductivity of PEDOT:PSS can significantly be increased by addition of secondary dopants such as dimethyl sulfoxide, ethylene glycol, or glycerol. Conductive silver inks are available from very fluid inkjet inks to very viscous pastes for screen printing. Inkjet inks are typically a colloidal silver nanoparticle suspension. The highly customizable, fine resolution printed patterns, grids, and electrodes require thermal sintering for conductivity. Thicker inks and pastes, on the other hand, consist of micrometer sized silver flakes. These are usually used for thick film electronics applications, where print layer reproducibility and bulk silver conductivity are of utmost importance.

Our study concentrates on using flexographical printing with conductive inks on various coated paper grades to form large area conductive surfaces. We use two different natural fibre based substrates: a multilayer pigment coated grade and a nanoparticle coated paperboard in comparison to traditional plastic film. The paperboard surface is functionalized by TiO_2 nanoparticles using a liquid flame spray (LFS) coating process. In LFS process liquid precursor of titanium (IV) isopropoxide

(TTIP) is fed into a high temperature and velocity flame in which the metal salt evaporates and nucleates to form nanoparticles of the metal oxide. These nanoparticles can be collected on a paperboard surface in on-line process flow and they cover the whole surface passing under the flame. For details of the LFS process, see Ref. [10].

It is known [11] that ultraviolet (UV) illumination can be used to change the wettability of TiO_2 surfaces, and similar photocatalytic wettability conversion has been observed on TiO_2 nanoparticle coated paperboard surfaces from superhydrophobic (water contact angle (CA) over 150°) to superhydrophilic (water CA less than 20°). Typical coating amounts on surface by the LFS are between 10 and 50 mg/m^2 [10] being several orders of magnitude smaller than in conventional dispersion coating concepts.

Here we report the effects of coating structure and surface wettability on formation of conductive surfaces. Tunable wettability of TiO_2 nanoparticle coated substrates has an effect on printability with water-based inks such as PEDOT:PSS as ink setting is poorer on superhydrophobic surface that is also observed from measured optical densities of the printed layers. We also report significantly higher conductivities with Ag ink on multilayer coated paper than with PEDOT:PSS conductive polymer.

Materials and Methods

We use three different printing substrates in our study: a commercial poly(ethylene terephthalate) (PET) film (Mylar®A) is used as the reference plastic film, a multilayer coated paper, and a double pigment coated paperboard (200 g/m^2, Stora Enso, Skoghall, Sweden).

The multilayer coated paper contains a commercial finepaper (Lumipress 115, StoraEnso, FI) that is coated by a 10 g/m^2 barrier layer consisting of platy kaolin (Barrisurf HX, Imerys Minerals Ltd, UK) blended with 50 pph ethylylene acrylic latex (Aquaseal 2077, Paramelt B.V., NL). To adjust printability, a 5 g/m^2 topcoating layer consisting of 70 pph fine platy kaolin (Barrisurf FX, Imerys Minerals Ltd. UK) and 30 pph fine blocky kaolin (Alphatex, Imerys Minerals Ltd., UK) blended with 6 pph SB latex (Basonal 2020.5, Basf, DE) was coated on top of the barrier layer. The paper was further calendered three times through a softnip at a line load of 120 kN/m and a temperature of 70 C°.

The paperboard sample was coated by TiO_2 nanoparticles using a coating and lamination pilot line at the Tampere University of Technology (Tampere, Finland) with a 30 m/min web speed. TTIP was dissolved in isopropanol with a metal ion concentration of 11.9 mg/ml with a 29.5 ml/min feeding rate into the nozzle whose distance from the moving web was set to 15 cm. This results in a deposition mass of 12.7 mg/m^2 with approximately 415 TiO_2 nanoparticles/μm^2 [12]. The average diameter of the nanoparticles is between 40 and 80 nm.

Flexographical prints were carried out using a laboratory scale IGT GST 2 printability tester. The anilox cylinder had a cell angle of 45° with 40 lines/cm and volume of 20 ml/m^2. PEDOT/PSS (Clevios P HC V4) conductive polymer dispersed in water with viscosity of 100 - 350 mPas was tested at a print speed of 2 m/s. Solvent based silver ink (Creative Materials 125-06 Extremely Conductive Ink) with viscosity of 2 000 – 3 000 mPas was tested at a print speed of 0.5 m/s. The contact angles of the inks were analyzed using a contact angle goniometer (KSV CAM 200) and the UV-exposure of the TiO_2 coated surfaces was carried out using a Bluepoint 4 Ecocure source with an UVA filter. The print densities were measured using a Techkon optical densitometer. Finally, a digital multimeter (Keithley 2100) was used to characterize the conductivities between two hand-painted contacts of silver conductive paint (Electrolube).

Results and Discussion

Contact angles

The contact angles of water, PEDOT:PSS (solid content 1.0 – 1.4 %), and silver printing inks for flexography are summarized in Table 1. Water contact angles are taken after 1 s for each sample whereas longer times for drop settling were required for more viscous PEDOT:PSS droplets (1 s for multilayer, 10 s for LFS paperboard, and 60 s for other samples). With TiO_2 nanoparticle coated samples we observe switching of superhydrophobic surface into superhydrophilic by UV light irradiation. Similar behavior is observed with water based PEDOT:PSS ink but no effect on silver ink is found. The LFS paperboard sample denotes the double coated paperboard without the TiO_2 nanoparticles.

Table 1. Measured contact angles of water, PEDOT:PSS ink, and silver ink.

	Plastic	Multilayer	LFS TiO_2	LFS TiO_2+UV	LFS Paperboard
H_2O	75	50	153	7	74
PEDOT:PSS	74	62	160	26	69
Ag	27	54	52	50	54

PEDOT:PSS prints

Optical print densities of PEDOT:PSS prints were measured before drying and are given in Table 2. The values for print layers are calculated by subtracting the measured background value for the unprinted substrate by $D_{print} = D_{measured} - D_{unprinted}$. As expected, the print densities increase after each successive print. Furthermore, the smallest values are observed for superhydrophobic surface indicating a poor transfer of ink to the substrate.

Table 2. Measured print densities of PEDOT-PSS before curing at the oven.

	Plastic	Multilayer	LFS TiO_2	LFS TiO_2+UV	LFS Paperboard
1st Print layer	0.02	0.03	0.01	0.07	0.01
2nd Print layer	0.06	0.06	0.04	0.09	0.03
3rd Print layer	0.08	0.09	0.06	0.13	0.09
4th Print layer	0.09				

Table 3 summarizes the measured conductance values with PEDOT:PSS flexographical prints after drying in an oven (1h, 120 °C). It was observed that plastic films require more successive print layers (here 4 layers) to obtain conductance due to poor wetting of smooth surface. The best performance was observed with a multilayer coated paper with controlled ink setting on the porous top layer. With TiO_2 nanoparticle coated samples we observe that surface wettability plays an important role when printing with water based flexographical inks. The conductivity values are lower when printing on superhydrophobic surface that is expected due to poorer wetting of the surface and formation of an uneven print layer. Hence, we can conclude that the superhydrophobicity survives even in the dynamic printing process and is lost only after several successive print layers.

Table 3. Observed sheet resistance of different substrates with PEDOT-PSS.

	Plastic	Multilayer	LFS TiO_2	LFS TiO_2 + UV	LFS Paperboard
1st Print layer	> 1000 MΩ/□	640 kΩ/□	> 1000 MΩ/□	1200 kΩ/□	3570 kΩ/□
2nd Print layer	> 1000 MΩ/□	200 kΩ/□	5250 kΩ/□	290 kΩ/□	270 kΩ/□
3rd Print layer	> 1000 MΩ/□	40 kΩ/□	330 kΩ/□	100 kΩ/□	80 kΩ/□
4th Print layer	620 kΩ/□				

It is worth emphasizing that the substrates here are not optimized for printing conductive layers, which explains the observed low conductivity (= high sheet resistance) values. For example, the LFS paperboard sample is a conventional paperboard used for packaging applications. With multilayer coating one can control ink setting resulting in better conductivities. Moreover, successive prints yield a more even and homogeneous print layer with higher conductivities [13]. The observed values are in agreement with the results shown by Barr *et al.* [6] with oCVD coated thin PEDOT layers.

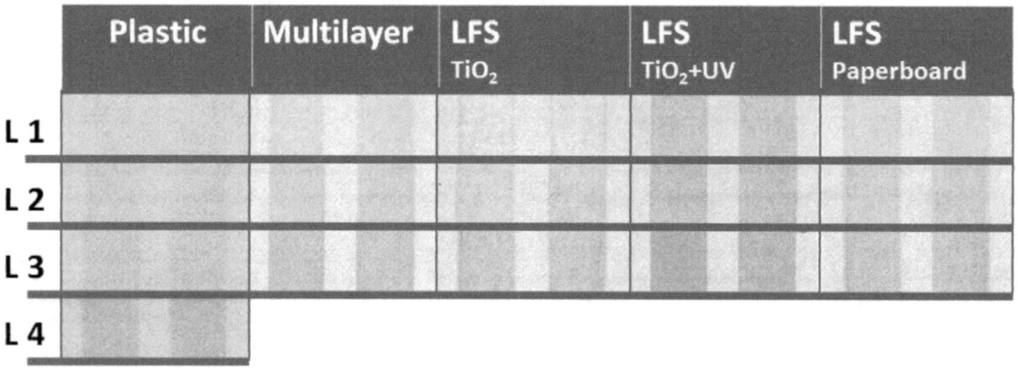

Figure 1. Flexographical PEDOT:PSS prints on different substrates (width of the printed area 7 mm).

Figure 1 displays the scanned images of the flexographical PEDOT:PSS prints. Poor ink setting both on plastic film as well as on superhydrophobic LFS TiO_2 substrate is observed that correlates well with the observed low conductivities. In multilayer coated structure the barrier layer below the thin and porous top coating layer prevents the ink penetration deep into the structure resulting in the highest conductance values.

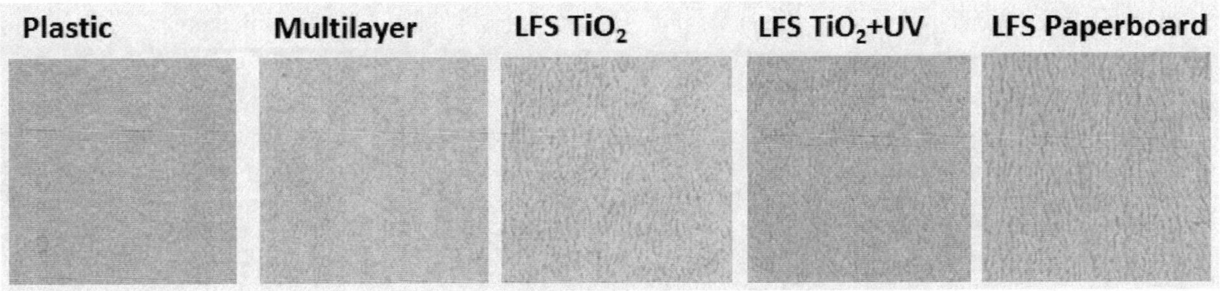

Figure 2. A larger magnification of the flexographical print layer (area 4×4 mm^2, 3 print layers).

Figure 2 displays a magnification of the printed layer structures. We observe a formation of branched clusters of polymer (dark areas), which are especially visible in the LFS paperboard reference sample. Similar behavior has been reported with spin-coated PEDOT:PSS thin films [14] that results in anisotropy in conductivity values in machine and cross-machine direction. This phenomenon was studied with contact angle goniometer as shown in Fig. 3. It is observed that with larger PEDOT:PSS droplets a neck is formed on the superhydrophobic TiO_2 LFS surface as the polymer concentrates on the droplet/substrate boundary. Initially the droplets behave like water droplets but as the polymer settles into the surface, the contact angle grows again with water minimizing the surface energy. The effect is rather slow for the large droplets. However, in flexography the size of the droplets is much smaller and hence, similar behavior may be observed in the print morphology.

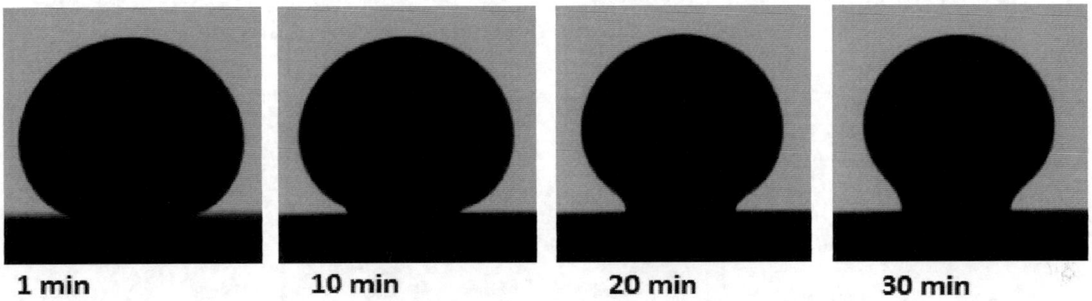

Figure 3. Time series of the PEDOT:PSS droplet on LFS surface.

Ag prints

We have recently demonstrated UV sensor functionality on multilayer coated paper by using flexograhically printed silver electrodes with conductive polymers [15]. Here we concentrate on identifying the properties of the flexographical process to improve conductivity of the prints. All silver samples were printed with 0.5 m/s printing speed. The pressure between the anilox cylinder and the printing plate as well as the pressure between the printing plate and the substrate was set to 100 N. Figure 4 shows microscopic images of three successive prints on multilayer coated paper with their corresponding resistance values. All samples were sintered in an oven at 120°C for 60 min. We observe non-conducting layer with a single print. This is due to two factors: first, the amount of ink transferred from the anilox is too small in order to fill the whole surface. Secondly, the number of cells/cm in our anilox cylinder may be too small compared to the cell volume *i.e.* individual cells are visible in the final print. Hence, no conductivity is observed as there is no connecting path between the individual silver flakes below the threshold value. After the second and third successive prints we observe good conductivity values as flakes become attached to each other.

Figure 4. Microscopic images of the flexographical Ag prints with 9 ml/m^2 and 100 lines/cm (magnification 100x).

The observed conductivity values are shown in Table 4. With each different gravure cell we observe non-conducting layer with a single print – this is expected as the amount of silver transferred to the substrate is too small to reach conductivity threshold. Similar behavior was observed in our earlier study [15] that related the surface coverage values to the observed conductivities.

Table 4. Observed sheet resistance of multilayer substrate with Ag ink using different cell sizes and line counts in the gravure cylinder.

	7 ml/m^2 160 lines/cm	9 ml/m^2 100 lines/cm	11 ml/m^2 80 lines/cm	16 ml/m^2 60 lines/cm	20 ml/m^2 40 lines/cm
1st layer	> 100 MΩ	> 100 MΩ	> 100 MΩ	> 100 MΩ	> 100 MΩ
2nd layer	130 – 156 Ω	35 Ω	1,1 – 1,3 MΩ	14 Ω	126 - 132 Ω
3rd layer	9,3 Ω	6,7 Ω	13 – 15 Ω	9 – 40 Ω	7,5 Ω

As a conclusion we have observed significantly higher conductivities with silver ink than with PEDOT:PSS. The reason for observing conductivity only after two successive prints is most likely due to the metallic anilox roller of our IGT GST2 printing unit, and the need for several successive prints for conductive electrodes is rather ineffective for practical applications. We believe that conductive layers with single print can be obtained by optimizing the printing conditions.

Conclusions

As a summary we have investigated possibilities to form large area conductive surfaces using a flexographical printing process with a laboratory scale test printer. Surface properties play a crucial role in the ink setting on the surface and in observed conductivity values. We have also shown that superhydrophobic surfaces result in poorer conductivity values with water based inks; to observe similar conductivity values as with hydrophilic surfaces several more prints are required. Furthermore, we have explored the optimal performance area for IGT printer with conductive inks: for water based

PEDOT:PSS with viscosity of 100 – 350 mPas faster printing speeds (>1.0 m/s) are required for better reproduction whereas for silver ink lower printing speeds (<1.0 m/s) are optimal. The lower speeds are needed for silver ink due to Ag flakes.

Comparing the observed conductivities between paper substrates to commercial plastic film for PEDOT:PSS we can conclude that the porous paper surface provides a more uniform ink setting, and hence, improved conductivity values. Silver ink provides several orders of magnitude better conductivities but they cannot be used as the semi-transparent top electrode. Further work is needed to optimize the coating structure and ink formulation to achieve high conductivities.

Acknowledgments

This work has been funded by the Academy of Finland (grant no 250 122 & 256 263). JJS wishes to thank the Japan Society for the Promotion of Science (JSPS) for an invitation to perform research at the Hokkai-Gakuen University in Sapporo, Japan in 2011 – 2012 under the "FY2011 JSPS Invitation Fellowship Program for Research in Japan (Long-term)" (grant no L–11537). We wish to acknowledge Imerys Minerals Ltd., UK, Paramelt B.V., NL, StoraEnso, FI, and Basf, DE for providing materials used in the multilayer coated paper.

References

1. Das, R. and Harrop, P. (2008), *Printed & organic electronics forecasts, players & opportunities 2008-2028*, IDTechEx report 7.
2. Tobjörk, D. and Österbacka, R. (2011), Adv. Mater. 23, 1935.
3. Bollström, R., Määttänen, A., Tobjörk, D., Ihalainen, P., Kaihovirta, N., Österbacka, R., Peltonen, J. and Toivakka, M. (2009), Org. Electron. 10, 1020.
4. Siegel, A.C., Phillips, S.T., Wiley, B.J. and Whitesides, G.M. (2009), Lab Chip 9, 2775.
5. Eveleigh, D. E. (1987), Phil. Trans. R. Soc. Lond. A 321, 435.
6. Barr, M. C., Rowehl, J. A., Lunt, R. R., Xu, J., Wang, A., Boyce, C. M., Im, S. G., Bulovic, V. and Gleason, K. K. (2011), Adv. Mater. 23, 3500.
7. Hübler, A., Trnovec, B., Zillger, T., Ali, M., Wetzold, N., Mingebach, M., Wagenpfahl, A., Deibel, C. and Dyakonov, V. (2011), Adv. Energy Mat. 1, 1018.
8. Berggren, M., Nilsson, D. and Robinson, N.D. (2007)., Nature Mater. 6, 3
9. Vosgueritchian, M., Lipomi, D. J. and Bao, Z. (2011), Adv. Funct. Mat. 22, 421.
10. Mäkelä, J. M., Aromaa, M., Teisala, H., Tuominen, M., Stepien, M., Saarinen, J. J., Toivakka, M. and Kuusipalo, J. (2011), Aerosol Sci. Technol. 45, 817.
11. Caputo, G., Nobile, C., Kipp, T., Blasi, L., Grillo, V., Carlino, E., Manna, L., Cingolani, R., Cozzoli, P. D. and Athanassiou, A. J. (2008), Phys. Chem. C 112, 701.
12. Stepien, M., Saarinen, J. J., Teisala, H., Tuominen, M., Aromaa, M., Kuusipalo, J., Mäkelä, J. M. and Toivakka, M. (2011), Appl. Surf. Sci. 257, 1911.
13. Denneulin, A., Blayo, A., Bras, J. and Neuman, C. (2008), Prog, Org. Coatings 63, 87.
14. Nardes, A. M. (2007), *On the conductivity of PEDOT:PSS thin films*, PhD Thesis, Technische Universiteit Eindhoven, The Netherlands.
15. Saarinen, J. J., Ihalainen, P., Määttänen, A., Bollström, R. and Peltonen, J. (2011), Nord. Pulp Paper Res. J. 26, 133.

(219d) Soybean Oil Based Fibers Formed without Solvent or Heat

Dustin W. Janes, Kadhiravan Shanmuganathan, Daniel Y. Chou and Christopher J. Ellison, Department of Chemical Engineering, The University of Texas at Austin, Austin, TX

The vast majority of commercially produced synthetic polymers used for fiber applications are made entirely from non-renewable, petroleum-based feedstocks. Because annual worldwide production of nonwoven fibers is in the billions of kilograms, incorporating even small amounts of biorenewable materials in these products could significantly impact the allocation of non-renewable resources. Additionally, large amounts of solvent and/or thermal energy are usually needed to process polymers into fibers. A safer and more efficient route to fiber production may be to photopolymerize functional monomer liquids during fiber formation processes. In this presentation we show that thiol/-ene chemistry can be harnessed to enable production of fibers containing over 50 wt. % acrylated epoxidized soybean oil without using solvent or heat. In this demonstration, the fibers were made by simultaneous electrospinning and photocuring of a liquid monomer mixture which we believe can be translated to other fiber manufacturing processes such as melt blowing. Scanning electron micrographs illustrate the fiber quality and an average diameter of about 30 μm. Photochemical conversion kinetics of functional groups during light exposure were measured by real-time Fourier transform infrared spectroscopy, providing insight into the advantages of using high-functionality monomers and thiol/-ene chemistry in this application.

VALIDATION OF CALENDERING MODELS FOR COATED PAPERS MADE FROM BLEND OF AGRI-RESIDUE AND SOFTWOOD PULP WITH SOFTNIP CALENDERING TECHNIQUES

A. K. Ray, *Department of Paper Technology, Indian Institute of Technology, Roorkee, Saharanpur, India*

Sanjay Tyagi, *Paper Testing and Energy management Division, Central Pulp and Paper Research Institute, Saharanpur, India*

ABSTRACT

Calendering is the final operation in the manufacturing of papers with the purpose for quality products by densification, and to increase smoothness (or to decrease the surface roughness) and gloss. In order to do so it decreases the thickness and bulk of paper. It is especially important in case of coated papers or paper boards where increase of gloss is essential. Paper is therefore, pigment coated to obtain a surface with improved optical and print performance additionally. However this is done under many modified operations such as supercalendering or softnip calendaring process. Generally, during calendering operation paper is subjected to compressive pressure pulse under the nip. After the nip, there is time dependent partial recovery of the initial thickness. The behaviour of the paper under calender nip is therefore viscoelastic in nature.

In the present paper modelling of stress – strain behaviour of coated paper under soft nip calender and its viscoelastic properties are attempted. The base paper used in this study was collected from the paper mill in reel form and having the furnish composition of 75% agricultural residue (bagasse/wheat straw) and 25% of soft wood chemical pulp. The paper coating trials were conducted on DT pilot coater at the speed of 20- 30m/min. During the run only one top side of paper was coated with the coat weight ranging 10 - 12 g/m^2. The coating formulation consisted of 100 parts pigment (fine ground calcium carbonate and fine clay in ratio of 80:20), 12 parts SB latex, 1 part CMC, 0.12 part OBA and 0.5 part dispersant). Calendering was carried out using soft nip rolls under varying calendering conditions of temperature from 60ºC to 85¢ªC and nip load of 84 kN/m to 120 kN/m. The moisture content of paper varies from 4.1% to 5.7 %. For drying of paper two IRT- Monocassettes of infrared units were used to give a temperature of 60 – 70°C. The temperature of air drying unit was kept at 50 °C .The data obtained after softcalendering were subjected to various investigation.

Available models for viscoelastic properties were critically examined and systematic estimation of the interrelated parameters of the viscoelastic models were made. The models were simulated with the data obtained from laboratory soft nip calender for coated paper. The flow chart is drawn to estimate the various parameters to the model simulation. The parametric effects were interpreted through various graphs and figures.

Key Words: Calendering, Softnip, Web temperature, Elastic modulus, Viscoelastic, Bulk

Introduction

Paper is pigment coated to obtain a surface with improved optical and print performance. Calendering is a very important technique for obtaining the above properties. It enhances the surface smoothness as well as gloss of coated paper or paper board. The apparent calendering effect is the surface flatness, the permanent compression and densification of the paper web. The parameters which affect the quality of paper developed through the calendering process are web speed/ speed of roll, pressure or total load, dwell time, calender roll surface temperature, number of nips, nip width, roll diameter, initial web temperature, web moisture, calender roll surface condition and the roll material properties.

Paper can be defined as highly porous random structure of flexible fibers, filler and other particles and can be considered to be elastico- visco- plastic material. As the paper web passes in to the converging nip, it experiences rapid compressive strain until the maximal compression is reached approximately at the nip centre. The main mechanism of paper under calender loading is the compressive deformation. Rodal (1989) pointed out in his model the viscoelastic behaviour of paper during calendering. The polymers in base paper and in the coating layers are either amorphous like lignin, hemicellulose and synthetic binder or partly crystalline like cellulose and starch. Deformation in the structure of polymers are time dependent and partly permanent (plastic) and partly recoverable (viscoelastic). The viscoelastic behaviour of polymer depends on molecular weight and polymer structure as well as external stress, temperature and moisture content. The response of viscoelastic material to the applied stress can be seen a combination of fast, viscous deformation and slower, elastoplastic deformation.

Menard (1999) described the viscoelastic properties of material by storage modulus (E'), loss modulus (E'') and by tanδ, which is the ratio between loss modulus and storage modulus. Browne et. al. (1996, 1994) has studied the importance of viscoelsatic behaviour of paper in hardnip calendering. The viscoelastic behaviour of the paper can be seen as an exponential paper thickness recovery after the nip. At higher linear loads the total recovery is constant and the portion of permanent deformation increases. This means that the whole paper structure deformation is a combination of the immediate or delayed elastic recovery and the permanent defrormation. Browne has shown that at low nip pressure almost all of an- in-nip deformation is recoverable, and the permanent deformation remains low. With increasing nip pressure, the recoverable deformation decreases both as an absolute value as well as compared to permanent deformation. This means that paper shows pure elastic behaviour at low nip pressure, but the importance of plastic deformation increases with increasing nip pressure. In the present study an attempt has been made to validate the viscoelastic models by Browne et. al.

for coated paper, calendered soft nip under laboratory conditions. The results were simulated using the data gathered for coated paper

Models for Viscoelastic behaviour of paper

The common way illustrating the viscoelastic models also termed as rheological models, is to use simple mechanical models involving spring and dashpot elements (Popil, 1989). The spring element is used to describe the idealized elastic behavior and the dashpot for viscous effects. If the model involves only linear behavior, the same is considered as linear viscoelastic model with constant elasticity and viscous parameters. The most commonly used models are

1) Linear Maxwell model
2) Linear Kelvin model (Voight model)
3) Burgers four element model

Linear Maxwell Model: Connecting a spring and a dashpot serially Maxwell model is obtained which characterizes the viscoelastic deformation of fluids which represents the linear Maxwell model where E is the elastic coefficient of spring and η is the coefficient of viscosity.

Linear Kelvin model (Voight model): In Kelvin model spring and a dashpot have been connected in parallel. This model describes the viscoelastic deformation of solids (delayed elasticity).

Burgers model: The four element Burgers model consists of Maxwell component and a Kelvin component. This model accounts for many substantial effects observed in paper like instantaneous deformation and recovery, permanent deformation and delayed recovery.

Systematic Estimation of the interrelated parameters of the viscoelastic model:

The details of derivation of general viscoelastic model are given in reference (Tyagi, 2009). The major equations used for the development of algorithm are given in Table 1.

Experimental Investigation

Base Paper: The base paper used in this study was collected from the paper mill in reel form and having the furnish composition of 75% agricultural residue (bagasse/wheat straw) and 25% of soft wood chemical pulp.

Coating formulation and coating trials:

The paper coating trials were conducted on DT pilot coater at the speed of 30m/min. The distance between the applicator roll and blade was kept at 250mm. The metering blade measured 84 x 315 mm was used in coating trials. The blade angle was kept at

31°. The blade load was adjusted by pneumatic cylinders at the pressure of 1.5 bar. During the run only one top side of paper was coated by the coat weight of 10 - 12 g/m^2. The coating formulation consisted of 100 parts pigment (fine ground calcium carbonate and fine clay in ratio of 80:20), 12 parts SB latex, 1 part CMC, 0.12 part OBA and 0.5 part dispersant). For drying of paper two IRT- Monocassettes of infrared units were used to give a temperature of 60 – 70°C. The temperature of air drying unit was kept at 50 °C

Calendering Conditions

After coating the sheets were calendered in laboratory calender. For paper coated with different coating formulations calendering was carried using soft nip rolls under varying calendering conditions of temperature from 60ºC to 85˚C and nip load of 84 kN/m to 120 kN/m. The speed of calendar kept at 22 m/min. The moisture content of paper varies from 4.1% to 5.7 %.

Evaluation of coated papers

Calendered coated/uncoated sheets were conditioned at 27± 1% relative humidity of 65 ± 2% prior to testing. All testing was done according to relevant ISO standard test methods.

Simulation of models for stress-strain behaviour of coated paper under soft nip calendering

The calendering parameters such as roll diameter, roll temperature, nip load, speed, nip intensity and consequent paper bulk, average nip pressure, nip residence time for softnip calenders are given in Table 2.

Using the calendering conditions given in Table 2, experiments were conducted on paper coated in laboratory using pilot blade coater. The regression equations were developed on the model based on Gerspach Equation (1993) and are given below

$$0.400 = a_0 + 1.92a_L - 0.68a_\sigma + 0.64at + 60a_T + 5.7a_M$$
$$0.600 = a_0 + 1.95a_L - 0.66a_\sigma + 0.66a_t + 65a_T + 5.4a_M$$
$$0.280 = a_0 + 2a_L - 0.64a_\sigma + 0.68a_t + 70a_T + 5.1a_M$$
$$0.400 = a_0 + 2.04a_L - 0.621a_\sigma + 0.70a_t + 75a_T + 4.7a_M$$
$$0.333 = a_0 + 2.06a_L - 0.613a_\sigma + 0.70a_t + 80a_T + 4.2a_M$$
$$0.130 = a_0 + 2.08a_L - 0.602a_\sigma + 0.72a_t + 85a_T + 4.0a_M$$

The values of unknown parameters of the calendering equations were calculated through least square techniques so that the sum of the squares of the error is minimized. The values are given in below

$A = -0.102$, $a_0 = 0.2797$, $a_L = 0.2008$, $a_\sigma = 0.2321$, $a_t = 0.0654$, $a_T = -1.4350$

$a_M = -2.4957$

Base on equations 1, 6, 7 and 8, mentioned in Table 1, the relationship was found between the linear load, permanent strain and recoverable strain for coated paper. The in- nip length and stress at nip was also calculated. The values of viscoelastic modulus for coated paper were calculated using the equations 10, 11 and 12. Based on equations and experimental results an algorithm describing the procedure for estimating parameters of viscoelastic models was developed for simulation purpose given in Figure 1. This algorithm is of extreme importance in the development and design of new calenders and in proving their validity for any type of paper.
Based on above algorithm a computer programme was developed in C++ to simulate the above models of stress- strain behaviour of paper coated with the one formulation and calendered with soft nip.

Results and Discussion

The relationship between linear load and permanent strain, in- nip strain and recoverable strain for coated paper calendered under soft nip condition were depicted in Figs.2 – 3. The computer generated regression equations are also given in graphs along with the respective regression coefficients.

Simulation carried out on experimental data generated in laboratory soft nip calendar indicates that model parameters are not merely the material properties but are the function of process parameters. Figure 2 indicate that as linear load varies from 84kN/m to 120kN/m, stress induced in coated paper also increases resulting in a linear increase in recoverable strain. There is little deviation found in the values of permanent and in-nip strain. The model predicted is in excellent agreement with experimental data as R^2 values lies in the range of 0.98 to 0.99
Figure 3 to 6 indicated the relationship of Maxwell elastic and viscoelastic modules and Kelvin elastic and viscoelastic modular relationship with calender loads. Clearly, as the load varies both Maxwell and Kelvin elastic and viscoelastic modulus varies linearly. Again, it is well in accordance to the work done by Corotgino et. al. (8) in the case of hard nip calendering of non coated papers.

Conclusions

Available models for viscoelastic properties of coated paper were critically examined. Systematic estimation of the interrelated parameters of viscoelastic models such as in-nip strain and permanent strain and their relationship are made. Calendering equation for soft nip calendars due to Gerspach was compared for prediction of viscoelastic properties of coated paper. Calculation of in-nip and out-nip width of paper, the values of Kelvin pair G_k and η_k were done. In case of validation of viscoelastic models, results concluded that as the linear load varies the in-nip strain and permanent strain induced in coated paper also increases. Compared to these two the values of recoverable strain found more. Both in-nip and permanent strain exhibits linear relationship with load. The relationship between Maxwell and Kelvin elastic and viscoelastic modules also varies linearly with applied load.

Nomenclature

L	Line load (kN/m)
B_i, B_n, B_f	Initial, in-nip and final bulk (cm^3/g)
L_i, L_n, L_f	Initial, in- nip and final thickness (cm)
ϵ_n, ϵ_p, ϵ_r	In- nip, permanent and recoverable strain
a	In going nip length (cm)
b	Out going nip length (cm)
σ	Stress (kN/m^2)
A_n, a_{0n}, a_{Ln}, a_{sn}, a_{Rn}	Calendering coefficients for hard nip calenders (gm/cm^3)
A_0, a_l, a_σ , a_t, a_T, a_M	Calendering coefficients for soft nip calenders
μ	Nip intensity
S, V	Web speed ((m/min)
T_k	Time constant
R	Roll radius (m)
T	Temperature of web (°C)
M	Web moisture (%)
G_m	Maxwell elastic modulus (MPa)
η_m	Maxwell Viscosity (k Pa s)
G_k	Kelvin elastic modulus (M Pa)
ϵ_k	Kelvin Viscosity (K Pa s)

References

1. Rodal, J., Soft nip calendaring of paper and paper board, TAPPI J., 72 (1989) 5, 77 – 186.
2. Menard, K., Dynamic mechanical analysis – A practical Introduction, CRC Press LLC, 1999.
3. Browne, T. C., Crotogino, R. H., Douglas, W. J. M., Viscoelastic modeling of paper in a calender nip, JPPS Vol. 22, No. 5, May 1996, J170 – J173.
4. Browne, T. C., Crotogino, R. H., Douglas, W. J. M., Measurement of paper strain in the nip of an experimental calender, JPPS vol. 20, No. 9, 1994 J266 – J271.
5. Popil, R. E., The calendaring creep equation – a physical model, Fundamentals of Papermaking Transaction of the ninth fundamentals research Symposium, held at Cambridge and published by Mechanical engineering Publication Ltd., 1989.
6. Sanjay Tyagi, Response of coating and calendaring on quality of papers made from indigenous raw materials, Ph. D. Thesis, submitted to Dept. of paper Technology, IIT Roorkee (Dec 2009).
7. Gerspach, A., Wochenbl Papier 121: 76 (1993). (cf. Mikko Jokio, Papermaking Part 3, Finishing, Book 10, Papermaking Science and Technology, Finish Paper Engineers' Association and TAPPI, 1999)
8. Corotogino, R. H., "Machine calendaring – recent advances in theory and practices", Trans. Tech. Section, Canada P. P. A., 7 Dec, TR 75 (1982).

Table 1: Major equations used for predicting viscoelastic model of coated paper

Eq. No.	Parameter	Equation
1	In- nip strain	$\epsilon_n = \frac{L_i - L_n}{L_i} = \frac{B_i - B_n}{B_i}$
2	In-nip Strain predicted by Calendering Equation	$\epsilon_n = A_n + \mu_n B_i$ Where $\mu_n = a_{0n} + a_{Ln} \log_{10} L + a_{Sn} \log_{10} S + a_{Rn} \log_{10} R$ (for hardnip calenders)
3	Gerspach Equation (for soft nip calendars)	$\mu = a_o + al \log L + a_\sigma \log\sigma + a_t \log t + a_T T + a_M M$
4	Permanent Strain	$\epsilon_p = \frac{L_i - L_f}{L_i} = \frac{B_i - B_f}{B_i}$
5	Permanent Strain Predicted by Calendering Equation	$\epsilon_p = A_p + \mu_p B_i$ Where $\mu_p = a_{0p} + a_{Lp} \log_{10} L + a_{Sp} \log_{10} S + a_{Rp} \log_{10} R$ and B_i is in range $-\left(\frac{A_p}{\mu}\right) \leq B_i \leq \frac{(1-A_p)}{2\mu}$
6	Strain predicted by Master Creep Equation	$\epsilon = A_p (1+\tanh \mu)$
7	Relationship between in nip and permanent strain	$\epsilon_n = C_0 + C_1 \log_{10} \epsilon p$
8	In- going nip	$a^2 = 2RLi\, \varepsilon_n$
9	Recoverable strain	$\varepsilon_r = \varepsilon_n + \varepsilon_p$
10	Kelvin pair G_k	$G_k = \frac{L}{V\tau_k(\varepsilon_R - \varepsilon_{el})}$
11	Strain	$\eta_m = L/(V.\, \varepsilon_p)$
12	Kelvin pair G_k and η_k	$\eta_k = \frac{L}{[V(\varepsilon_R - \varepsilon_{el})]}$
13	Van Haag Model	$\sigma_z = (E_0\, \varepsilon_z)/(1 - \varepsilon_z/\varepsilon_{max})$

Table 2: Calendering conditions for soft nip calendering of coated paper

Roll Temp ,°C	60	65	70	75	80	85
Nip dwell time t (ms)	4.4	4.54	4.79	5.02	5.13	5.24
Nip load (kN/m)	84	90	100	110	115	120
Nip pressure (kN/mm^2) (σ)	0.208	0.216	0.228	0.239	0.244	0.250
Nip width (mm)	1.61	1.66	1.76	1.840	1.88	1.92
Bulk (cm^3/g)	1.12	1.07	1.00	0.95	0.92	0.88
Nip intensity (μ)	0.448756	0.458577	0.473215	0.486276	0.492239	0.498399

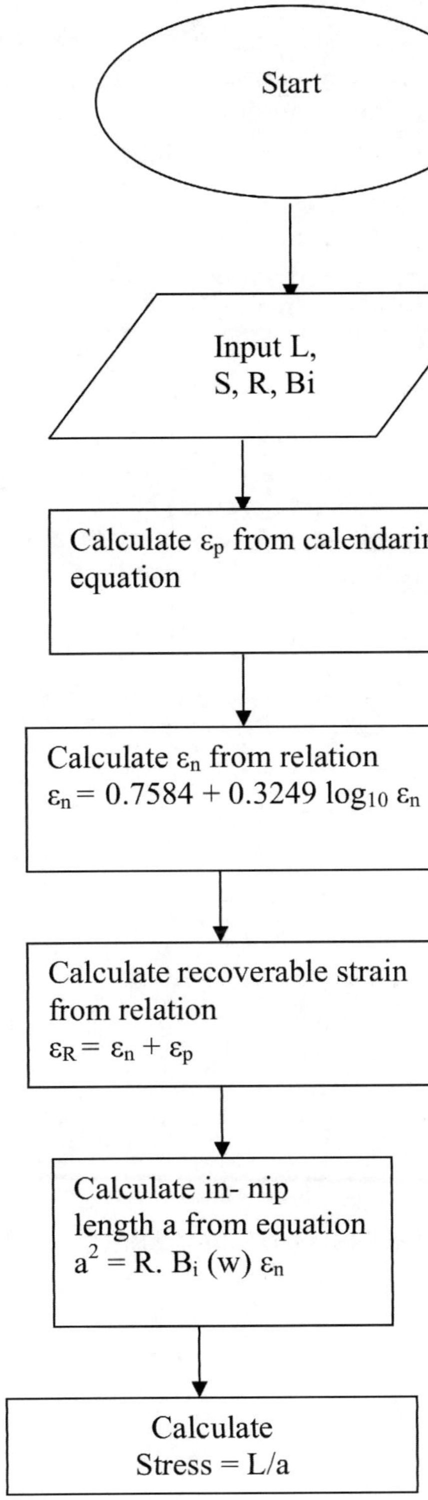
Start
Input L, S, R, Bi
Calculate ε_p from calendaring equation
Calculate ε_n from relation $\varepsilon_n = 0.7584 + 0.3249 \log_{10} \varepsilon_n$
Calculate recoverable strain from relation $\varepsilon_R = \varepsilon_n + \varepsilon_p$
Calculate in- nip length a from equation $a^2 = R.\ B_i\ (w)\ \varepsilon_n$
Calculate Stress = L/a

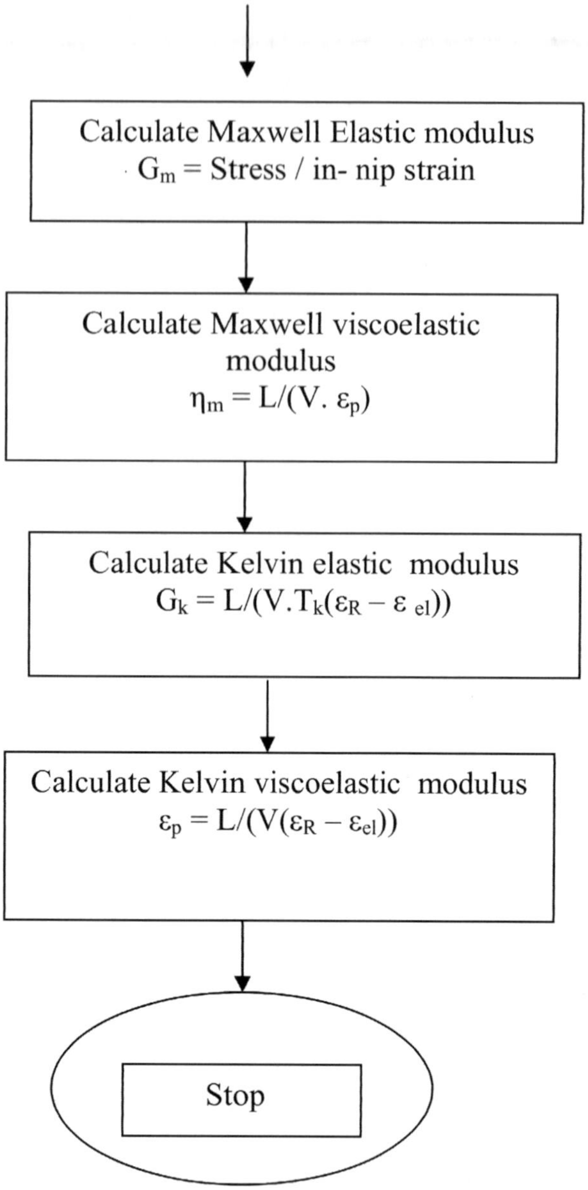

Figure 1: Procedure for estimating parameters of viscoelastic model

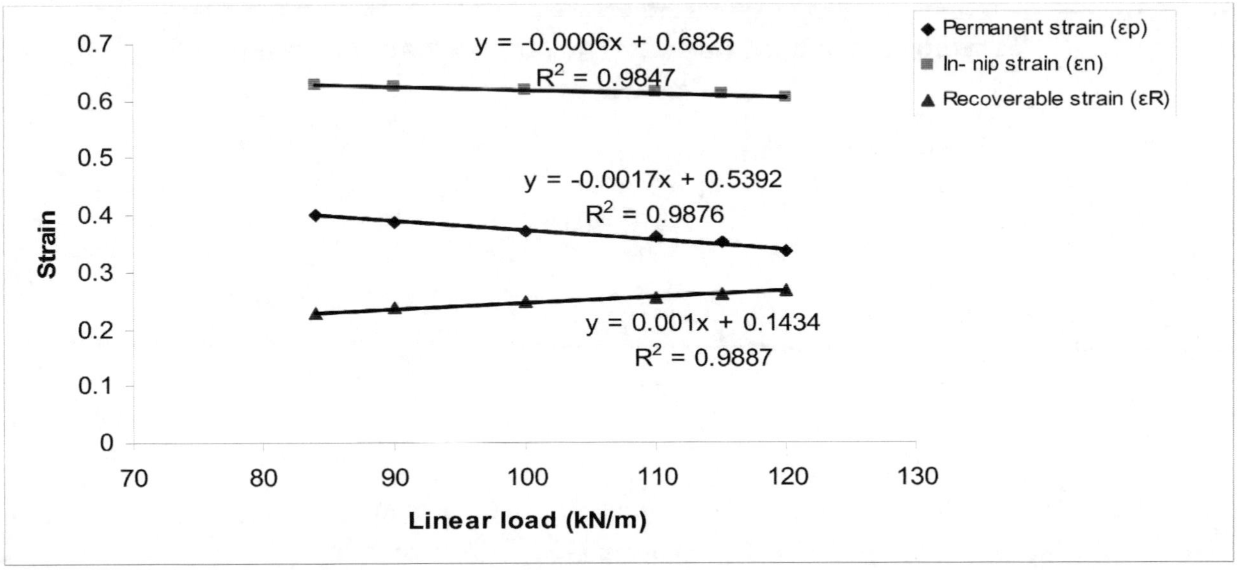

Fig 2: Permanent strain, in- nip strain and recoverable strain of coated paper with varying load (softnip calender)

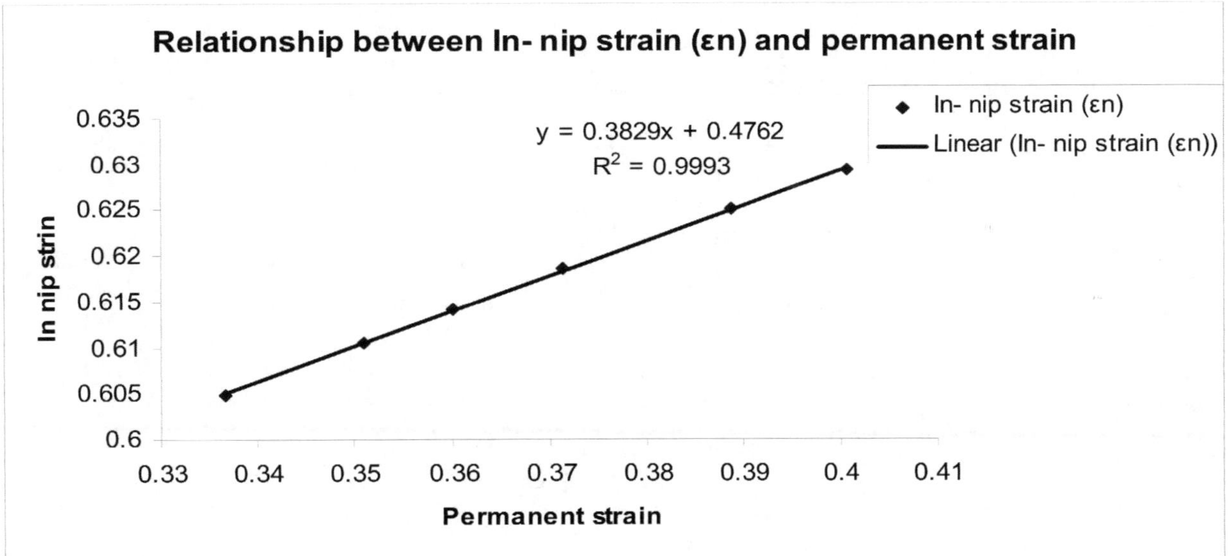

Fig. 3: Relationship between Permanent strains and In- nip strain for coated paper (soft nip calender)

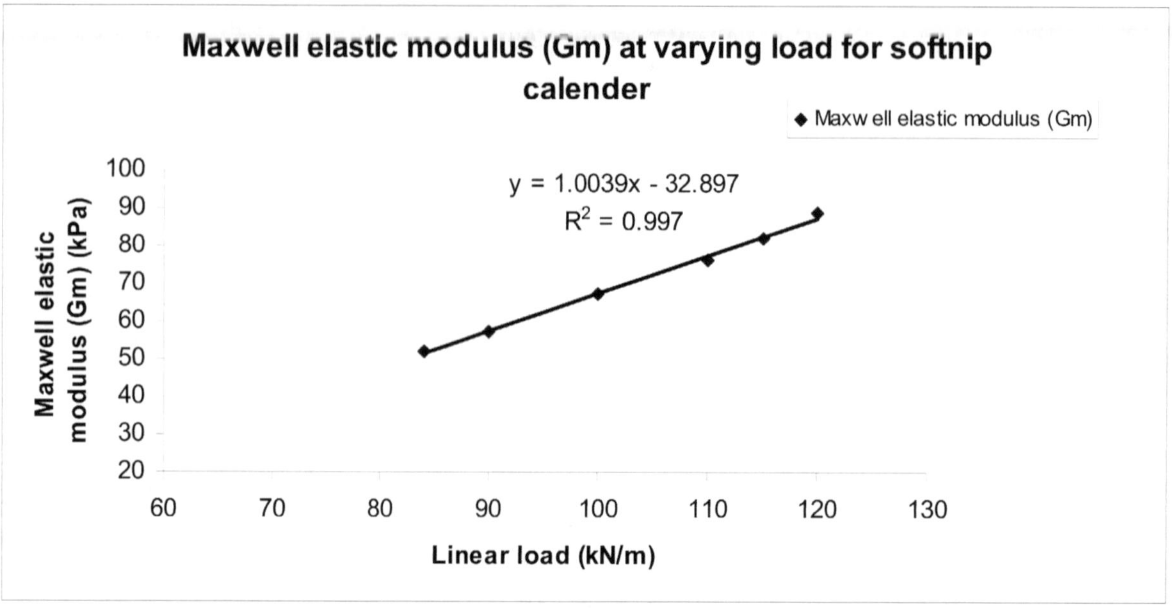

Fig. 4: Maxwell elastic modulus (G_m) of coated paper at varying load (softnip calender)

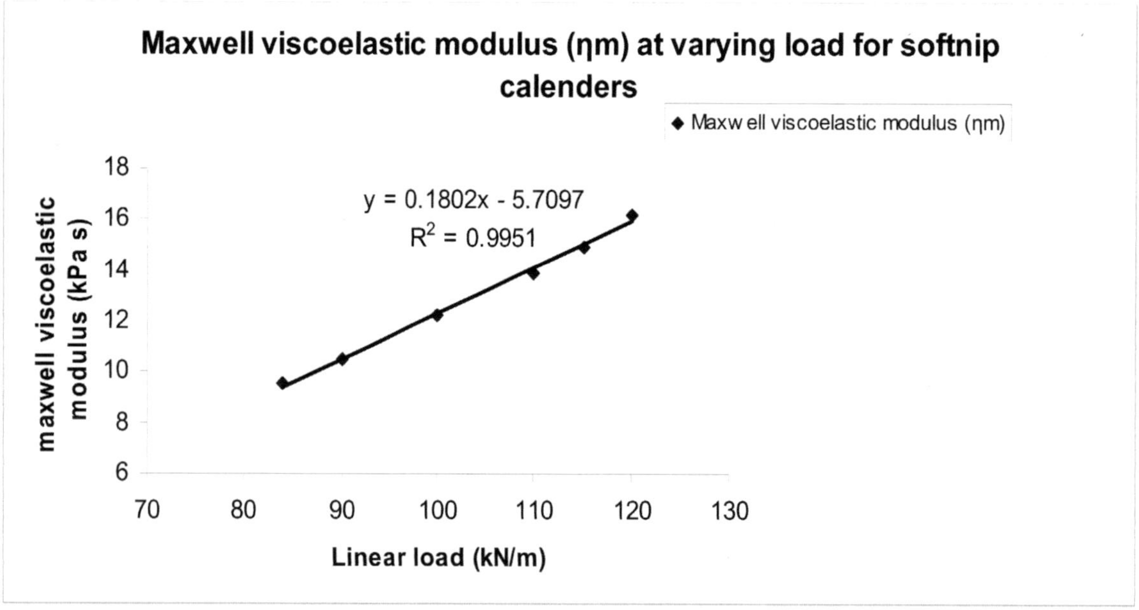

Fig 5: Maxwell viscoelastic modulus (η_m) of coated paper at varying load (softnip calender)

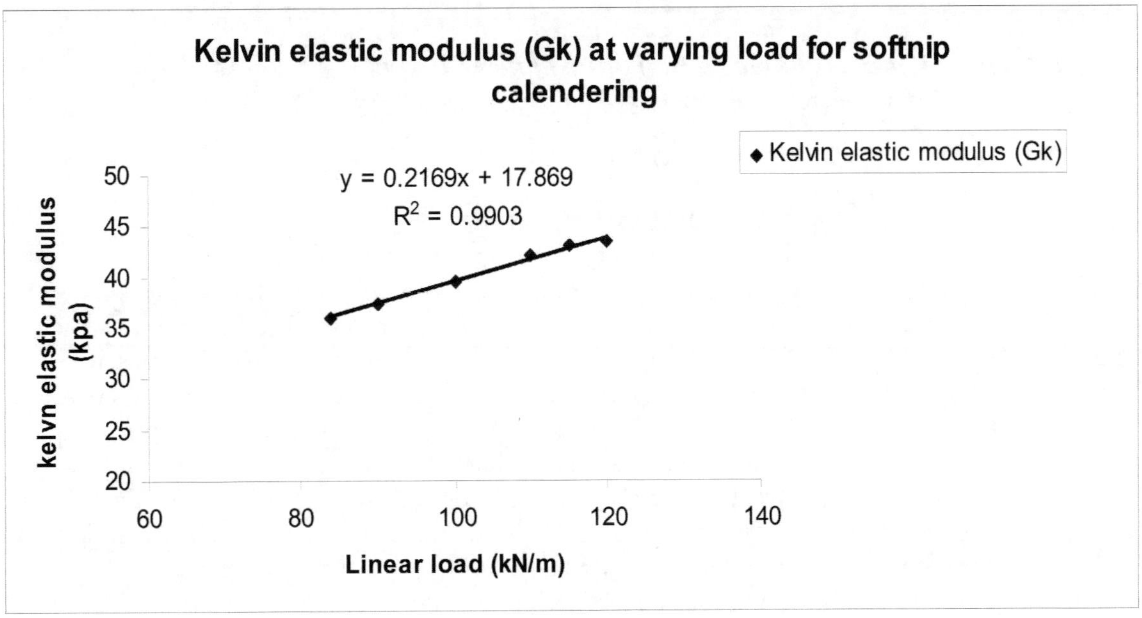

Fig 6: Kelvin elastic modulus (G_k) of coated paper at varying load (softnip calender)

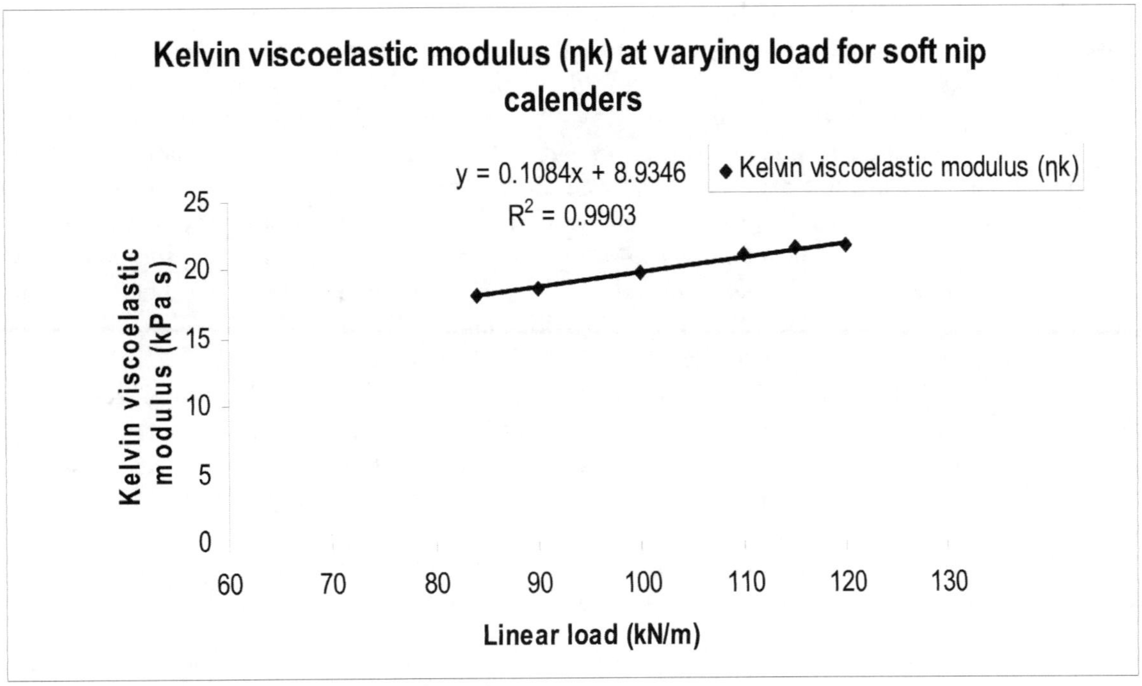

Fig 7: Kelvin viscoelastic modulus (ε_k) of coated paper at varying load (softnip calender)

Characterization of sludge from combined deinking technology

Shilpa Kulkarni[1a], Vivek Kumar[1] and M. C. Bansal[1],
[1]Department of Paper Technology, Indian Institute of Technology- Roorkee, India
Corresponding author: [a]*shilpa.ips@gmail.com*

1. INTRODUCTION

Recycled fiber has become an important source for the production of the paper. In the early 70's decades, only 7% of waste paper was used as a raw material for the production of the paper which is now increased up to the 47% and acting as a major raw material for paper industry. It is possible due to improved recycling system of wastepaper. Presently almost all grades of paper are being recycled. The share of mixed office waste (MOW) in the recovered wastepaper has increased rapidly during last decade. Non-impact printed paper represents the major share in MOW.

The deinking process is a process which is mainly utilized for removal of paper ink .The sludge generated during this process is called deinking sludge. Its disposal is one of the most important tasks for environmental protection. In recent years, deinking mills have striven to put their sludge by products to positive use [1-3]. The detail characterization of deinking sludge provides important information such as reutilization and proper treatment of sludge before disposal. However, the characteristics of sludge produced by the mill depend on the type of the wastepaper and deinking process. The sludge largely comprise of ink waste, generally with 60% moisture content and varying amounts of Si, Al, Cu, Ti, Fe and Ca as well as organic matters [4-6] and these by-product largely sent to landfill.

The most common options for the processing of paper industry sludge range from their exploitation for agricultural purposes, compositing, or use as a primary material in the manufacture of ceramics and cement [7-9], to energy recovery in biomass boilers or fluidized bed systems.

Each deinking operation produces sludge with somewhat unique characteristics, depending on the type of waste paper processed and deinking process. Waste paper with high amounts of filler, such as clay-coated magazine paper, produces much lower yields of usable fiber than waste paper having most of cellulose fiber, such as computer printout or old newspaper. Lightly printed paper produces less sludge than the heavily printed paper. Similarly alkaline papers contain high filler-to-fiber ratios and, therefore, produce more sludge. In present scenario different deinking process are used. Enzymatic deinking is advantageous for industrial usage because it is efficient, quick and has a low environment impact. Besides, UV-irradiation and Ultrasound technologies have begun to be investigated by some researcher, which is still in its starting stage [10-13]

Characterization of denking sludge generated from chemical deinking process is well reported in literature [12]. Various techniques can be applied to characterize the sludge [6, 11]. The present work focuses on the characterization of deinking sludge generated from photocopier waste paper by utilising various chemical and combined deinking treatment process namely: i) Chemical deinking, ii) Enzymatic deinking, iii) Combined deinking (Chemical + Enzyme + UV-irradiation and Chemical + Enzyme + Ultrasound) technology, iv) Chemical + UV-irradiation, v) Chemical + Ultrasound, vi) Enzyme + UV-irradiation and vii) Enzyme + Ultrasound treatment.Various researchers are investigated on deinking sludge but no result was found on the characterization of combined deinking sludge.

2 Material and Methods

2.1 Raw material

Photocopier grade paper (basis weight 70 gsm) was used to produce waste paper after printing with Xerox Black toner, using Xerox Workcenter pro420 photocopier machine. All sheets were uniformly printed.

2.2 Experimental plan :

Flow chart of experimental plan was show in Figure 1(a) chemical deinking and combined deinking and Figure 1(b) Enzymatic deinking and combined deinking

2.3 Deinking process

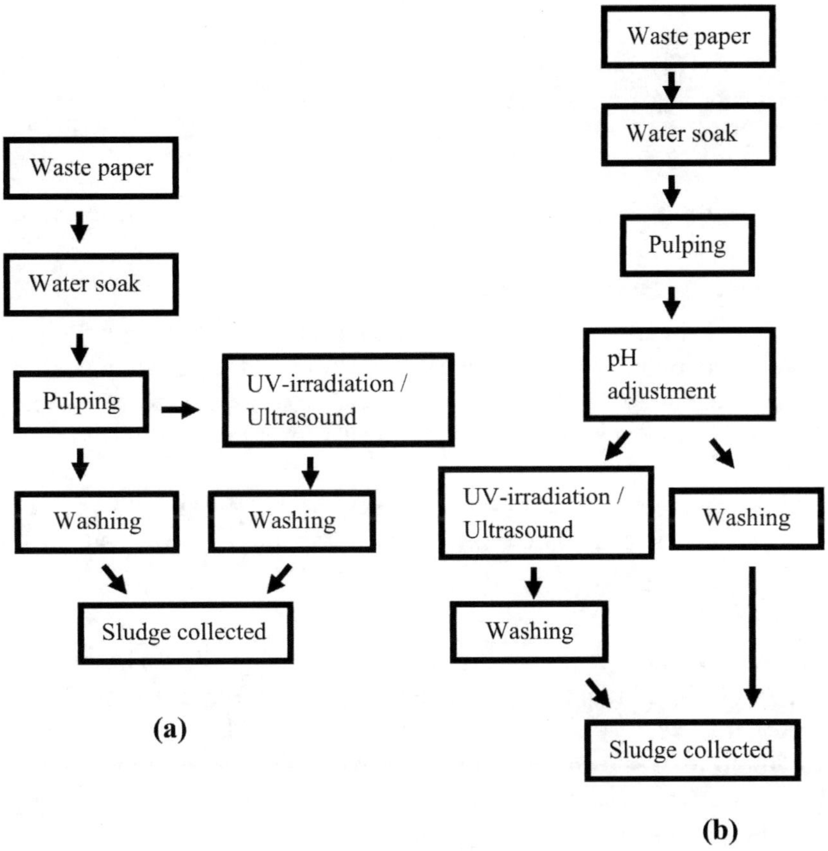

Figure: 1 (a) chemical deinking and combined deinking process **(b)** Enzymatic deinking and combined deinking process

2.3.1 Pulping

Hydrapulper is used for slushing of the xerographic paper used during the process under controlling temperature and at fixed rotor speed 180 rpm. The detail conditions for chemical dosing and enzymatic pulping are shown in **Table 1.**

Table 1: Pulping conditions and chemical dosing

Specification	Chemical deinking	Enzymatic deinking
Hydrapulper consistency	8%	8%
Pulping time of hydrapulper	25-30 min	25-30 min
Temperature	60ºC	55ºC
Chemicals	Sodium hydroxide- 2.0% Sodium silicate -2.5% Hydrogen peroxide- 0.2% DPTA -0.5%	-
Surfactant	Oleic acid -0.8%	Oleic acid-0.8%
Enzyme dosing	-	Cellulase-10 IU/g O.D paper
pH	-	4-5
Enzyme incubation	-	45 min
UV-irradiation time	25 min	25 min
Ultrasonication time	25 min 15 min	25 min 15 min
Washing time	5 min	5 min

2.3.2 Washing

Washing was carried out by using 60 mesh wire sieves for 5min (~20L water) under running tap water.

2.3.3 Enzyme

Commercial cellulase enzyme (Hi-media) was used for deinking purpose. The enzyme was stored at 4°C prior to use. This enzyme acts mainly on the surface of the fiber in the acidic to neutral pH range (pH 4.0 to 5.0) and between temperatures of 40 and 55°C. The cellulase activity was measured using the carboxymethyl cellulase (CMCase) method [14]. Its highest activity was found to be 1000 IU/g O.D paper at 55°C and pH 4.8. .

The enzyme dose were optimized from the range between (0-15) IU/g O.D paper from this optimization 10 IU/g O.D paper was best. The enzyme was dissolve in citrate buffer having pH (4.8).

2.3.4 UV –irradiation treatment

UV–irradiation treatment was carried out in a plastic container for 25 min.. The plastic container containing the pulp mixture was placed in a timber-framed UV (ultra violet) reactor (77cm ×36cm × 71cm) equipped with (λ =365nm), 72W (Philips) on the top side, located at a distance of 15cm from the sample. The reaction mixture was subjected to irradiation under UV light with continuous stirring. A fan fitted on the side wall was used to lower the heat generated by UV lamps. All the experiments were carried out in completely mixed and batch mode.

2.3.5 Ultrasound treatment

Ultrasound treatment was carried out in an Ultrasonicator (Model no. UTS-120, Unitech Ultrasonics, Mumbai, India) for 15 min and 25 min. The glass container containing the pulp mixture was placed inside chamber. The size of chamber is (150×150×150) in mm, chamber capacity is (3L), and Power of chamber is (120W)

Ultrasound which are converted at high efficiency into mechanical vibrations. This causes violent distension and contraction of cleaning fluids giving rise to millions of microscopic vacuum bubbles which forms and collapse at the same rate and produce a by-product known as "Cavitation " It produces intense scrubbing action throughout the liquid thereby rapidly and efficiently removing contaminants even from the intricate parts of the components.

2.4 Paper Sheet and Sludge characterization

Hand sheets were prepared by using Tappi test method T 205 sp-95. The tensile strength, tear index, and burst strength of paper sheet were determined by the Tappi test method T 494 om -96, T 414 om -98 and T 403 om-97 respectively. Sludge obtained by the washing of the waste paper was characterized by the Thermo gravimetric analyses (TGA) and CHNS analysis. TGA of deinking sludge were carried using EXSTAR TG/DTA 6300. The condition of thermal analysis was from the temperature ambient to 1000°C with the heating rate of 20°C/min under the flow of nitrogen at 200 ml/ min.

3. Result and Discussion

After pulping in presence of various doses of enzyme, we did the flotation of pulp and further determined the brightness of paper sheet which was prepared from enzyme deinked pulp. Table 2 represented the brightness before and after flotation in various doses enzyme deinked pulp. From Table 2 we observed that utilization of 10 IU/ g of enzyme does give the higher brightness. In other word we can say that 10IU/ g of enzyme does removed more ink in comparison to other doses of enzyme.

Table 2: Optimization of enzyme dosing in deinking process

Optimization of enzyme dosing in deinking process		
Optimization of enzyme dosing (IU/g O.D paper)	Brightness Before flotation	Brightness after flotation
0	69.70	76.10
5	73.20	77.40
10	76.90	81.70
15	77.00	81.70

After optimization of enzyme dose we again did the pulping in presence of 10 IU/ g of enzyme dose pulp (contain higher brightness and further treated the pulp under either only by UV irradiation or ultrasonication or by all three combined treatment i.e. chemical, enzyme and UV irradiation/ ultrasonication) . Three other deinking treatments were done separately which was (1) Chemical, (2) Chemical and UV irradiation and (3) Chemical and ultrasonication. The brightness of all the deinking treatments was represented in the Table 3. We observed from Table 3 that (i) combined deinking treatments chemical + enzyme + ultrasound (15 & 25 min.), chemical + US (25min.) and chemical + US (15min.) are better in comparison to other treatments due to higher brightness. (ii) According to brightness results, Ultrasonication was found to be more effective than the only enzyme treatment and enzyme + chemical treatment.

In the Table 4, it shows that tensile index is increases ~26.4 % when combined deinking technologies were used as compared to individual chemical, enzyme and UV-irradiation treatment. The Tear index was increased about to 8.5% in both cases i.e. enzyme with UV-Irradiation treatment for 25 min and enzyme with Ultrasound treatment for 25min. It was more effective then the chemical treatment.

Table 3: ISO Brightness of chemical, enzyme and combined deinking treatments

Particulars	Deinking Treatments									
	Chemical treatment				Enzyme treatment				Chemical + Enzyme + UV	Chemical + Enzyme + Ultrasound
	Chemical	Chemical + UV	Chemical + US		Enzyme	Enzyme + UV	Enzyme + Ultrasound			
		25 min	25 min	15 min		25 min	25 min	15 min	25 min	25 min
ISO Brightness (%)	74.3	74.3	75.1	74.3	73.4	74.8	74.9	75.4	68.8	75.7

Table 4: Physical properties of chemical, enzyme and combined deinking treatment

Particulars	Deinking Treatments									
	Chemical treatment				Enzyme treatment				Chemical + Enzyme + UV	Chemical + Enzyme + Ultrasound
	Chemical	Chemical + UV	Chemical + US		Enzyme	Enzyme + UV	Enzyme + Ultrasound			
		25 min	25 min	15 min		25 min	25 min	15 min	25 min	25 min
Tensile index (mN/g)	10.8	13.1	16.0	15.0	11.1	13.3	7.7	10.9	10.6	13.4
Tear index (mN.m^2/g)	4.6	6.3	4.4	4.7	5.4	6.9	5.3	3.7	4.3	4.7
Burst index (kPa.m^2/g)	0.4	0.4	0.5	0.4	0.3	0.4	0.3	0.3	0.3	0.4

3.1 Thermal characterization

Figure 2 is the graph of chemical + enzyme + US (25min) treated sludge sample. The first weight loss in all treatment is very small weight loss which must be due to the presence of the moisture content in the sludge. The combustion of cellulose to form volatile combustion products found in second weight loss in between 100°C-370°C while residual material can be obtained at temperature 900°C. This should be percent of ash. The added fillers % in sludge can be observed by % weight loss between the temperature range 370-750 °C. Table 5 represents the % of various contents obtained by TGA technique in combined deinking process. From the Table 5 we have found that fiber content in sludge sample was found higher (~55 %) when it was treated by chemical + US , Enzyme + US and only enzyme treatments methods.

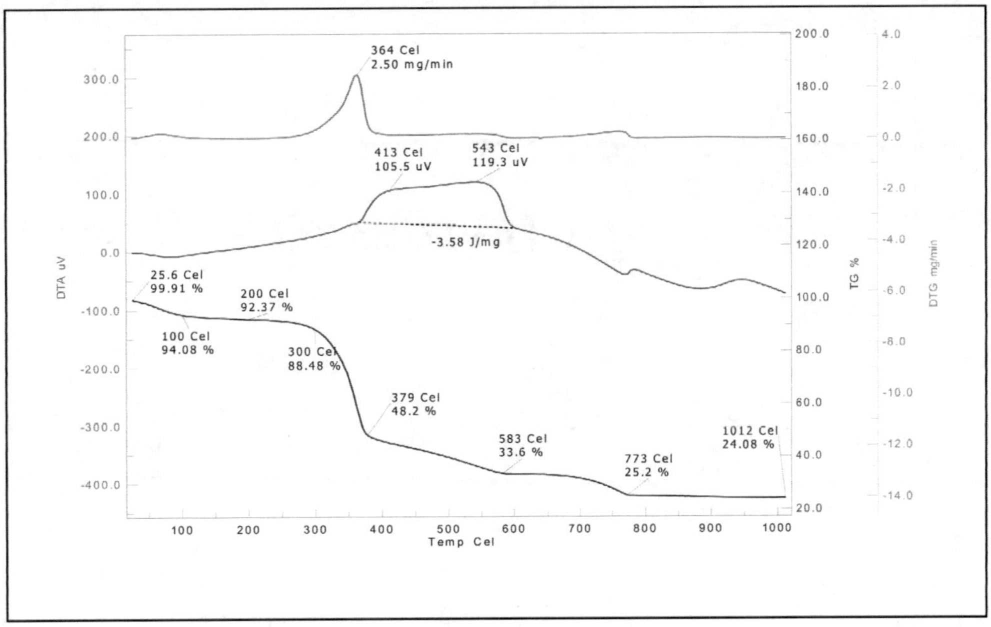

Figure 2: Thermal analysis of chemical + enzyme + US (25min) treated sludge sample

Table 5: TGA Analysis of chemical, enzyme and combined deinking treatment sludge

Percentage of various content on the basis of TGA analysis										
Particulars	**Chemical treatment**				**Enzyme treatment**				**Chemical + Enzyme + UV**	**Chemical + Enzyme + Ultrasound**
	Chemical	**Chemical + UV**	**Chemical + US**		**Enzyme**	**Enzyme + UV**	**Enzyme + Ultrasound**			
		25 min	25 min	15 min		25 min	25 min	15 min	25 min	25 min
Moisture % up to 100°C	4.12	49.10	4.14	6.10	4.09	-	6.80	6.29	6.15	5.90
Volatile combustion products% 100-370 °C	38.68	27.10	55.96	58.10	53.31	-	55.4	55.01	42.15	45.88
Inorganic % 370-750 °C	24.80	12.90	20.20	19.90	18.40	-	20.50	19.60	23.90	23.00
Residual Material at 1000 °C	31.36	10.31	18.67	15.21	23.21	-	16.34	18.01	27.19	24.08

3.2 CHNS Analysis

The CHNS Analyzer find utility in determining the percentages of Carbon, Hydrogen, Nitrogen, Sulphur and Oxygen of organic compounds, based on the principle of "Dumas method" which involves the complete and instantaneous oxidation of the sample by "flash combustion". Results are represented in Table 6. The results shows the higher % of carbon contents in sludge sample obtained by the chemical + US, Enzyme + US, Enzyme + UV and only enzyme treatment. It also confirmed that above said treatment contain higher fiber content in sludge sample.

Table 6: CHNS Analysis of deinking sludge

Particulars	Deinking Treatments									
	Chemical treatment				Enzyme treatment				Chemical + Enzyme + UV	Chemical + Enzyme + Ultrasound
	Chemical	Chemical + UV	Chemical + US		Enzyme	Enzyme + UV	Enzyme + Ultrasound			
		25 min	25 min	15 min		25 min	25 min	15 min	25 min	25 min
C	25.6	-	32.3	34.2	43.2	32.8	32.9	33.5	28.3	28.5
H	2.3	-	3.9	4.7	3.2	3.8	5.2	4.8	2.7	3.5
N	0.2	-	0.2	0.2	0.2	0.2	0.1	0.1	0.1	0.1

4. Conclusion

In denking process, the utilization of 10 IU/ g of enzyme doses removed more ink in comparison to other doses of enzyme. (1) Combined deinking treatments i.e. chemical + enzyme + ultrasound, chemical + US (25min.) and chemical + US (15min.) are better in comparison to other treatments due to higher brightness. Introduction of Ultrasonication treatment was found to be more effective in deinking process. Both tensile strength and Tear index is increases when combined deinking technologies were used as compared to individual chemical, enzyme and UV-irradiation treatment. Fiber content in sludge sample was found higher (~55 %) when it was treated by chemical + US, Enzyme + US and only enzyme treatments methods. In above result, it shows that enzyme treatment is effective and it improves the strength properties with combined technology

5. Acknowledgement

The authors wish to convey their sincere thanks to Dr R.M. Mathur, Director, CPPRI, and Saharanpur for their help and support.

6. References

[1] Kim S, Kim H-J, Park JC. Application of recycled paper sludge and biomass materials in manufacture of green composite pallet. Resources, Conservation and Recycling 2009;53:674-9.

[2] Lou R, Wu S, Lv G, Yang Q. Energy and resource utilization of deinking sludge pyrolysis. Applied Energy;2010; 90:46-50.

[3] Yan S, Sagoe-Crentsil K, Shapiro G. Reuse of de-inking sludge from wastepaper recycling in cement mortar products. Journal of Environmental Management; 2011; 92:2085-90.

[4] Pera J, Amrouz A. Development of Highly Reactive Metakaolin from Paper Sludge. Advanced Cement Based Materials 1998;7:49-56.

[5] Behin J, Vahed S. Effect of alkyl chain in alcohol deinking of recycled fibers by flotation process. Colloids and Surfaces A: Physicochemical and Engineering Aspects 2007;297:131-41.

[6] Méndez A, Fidalgo JM, Guerrero F, Gascó G. Characterization and pyrolysis behaviour of different paper mill waste materials. Journal of Analytical and Applied Pyrolysis 2009;86:66-73.

[7] Ahmadi B, Al-Khaja W. Utilization of paper waste sludge in the building construction industry. Resources, Conservation and Recycling 2001;32:105-13.

[8] Souza PSL, Dal Molin DCC. Viability of using calcined clays, from industrial by-products, as pozzolans of high reactivity. Cement and Concrete Research 2005;35:1993-8.

[9] Conesa JA, Gálvez A, Fullana A. Decomposition of paper wastes in presence of ceramics and cement raw material. Chemosphere 2008;72:306-11.

[10] Gaquere-Parker AC, Ahmed A, Isola T, Marong B, Shacklady C, Tchoua P. Temperature effect on an ultrasound-assisted paper de-inking process. Ultrasonics Sonochemistry 2009;16:698-703.
[11] Gea T, Artola A, SÃ¡nchez A. Composting of de-inking sludge from the recycled paper manufacturing industry. Bioresource Technology 2005;96:1161-7.
[12] Beauchamp CJ, Charest M-Hln, Gosselin A. Examination of environmental quality of raw and composting de-inking paper sludge. Chemosphere 2002;46:887-95.
[13] Zhenying S, Shijin D, Xuejun C, Yan G, Junfeng L, Hongyan W, et al. Combined de-inking technology applied on laser printed paper. Chemical Engineering and Processing: Process Intensification 2009;48:587-91.
[14] Ghosh TK. Measurement of Cellulase Activities. Pure & appli chem 1987;59:257-68.

POLYLACTIDE/CELLULOSE ACETATE/TRIETHYL CITRATE BLENDS AND THE EFFECT OF CELLULOSE ACETATE PERCENTAGE ON BLEND PROPERTIES

Barbara A. Wheelden, Amber R. Tupper, Sunggyu Lee,
Ohio University, Athens, OH

Abstract

Polylactide (PLA) is a bio-based, biodegradable polymer that is inexpensive and readily available. It is derived from lactic acid, which is produced via the fermentation of sugar-containing crops. Under suitable conditions, PLA will compost into the natural end products of carbon dioxide, water, and humus. Already used in a variety of compostable serviceware (cups, bowls, silverware, etc.) and other disposable products, PLA has gained popularity over the past few years as industry and the public have become aware of the impact of waste materials (especially non-degradable plastics) on the earth's ecosystems as well as the need for sustainable materials from renewable starting products. PLA's high modulus and low elongation at break, however, make the unmodified polymer unsuitable for uses in flexible packaging applications, an area where the need for biodegradable polymers is evident. In this study, PLA is blended with Natural Rubber (NR) to produce an all-renewable, eco-friendly polymer blend. Organoclay is used as a compatibilizer between the two immiscible polymers. Thermal, mechanical, and rheological properties of the blends at varying percentages of NR are studied.

Introduction

While PLA has a high tensile modulus, it is a very brittle material. Variations between commercial grades exist, but in general, PLA exhibits an elongation at break of approximately 5% [1], and a low impact strength of 47 kJ/m^2, a strength that is further reduced upon injection or compression molding [2]. Many methods have been utilized to improve these characteristics of the polymer. From a synthesis standpoint, co-polymerization of lactic acid and lactide has been carried out to produce copolymers of PLA and poly(glycolic acid) [1], polybutadiene [3], polyisoprene [4] and polycaprolactone [5]. Plasticizers utilized to enhance the flexibility of PLA include citrate esters and citrate ester oligomers [6, 7] and lower molecular weight polymers such as poly(ethylene glycol) and poly(propylene glycol) [8, 9]. Blending of PLA with polymers that possess desirable qualities is also a popular method for altering its mechanical properties. Because PLA is popular due to its renewability and degradability, most of the efforts to modify PLA have been carried out with materials that are either produced from renewable resources, are biodegradable, or both. Popular bio-based polymers which have been blended with PLA are polycaprolactone [10], polyhydroxyalkoanate [11], and poly(buylene succinate) [12]. PBAT, a petroleum-based, biodegradable polymer, has been shown to increase the elongation at break of PLA from 3.8% to >200%, when only 5 weight percent of PBAT was incorporated into the PLA matrix [13].

Recently, a popular method of toughening PLA has been melt blending the polymer with natural rubber [14-17]. Natural rubber is primarily composed of polyisoprene, which comes from latex. Latex is produced by over 20,000 species of plant [18], and of these, approximately 2500 species produce latex that is a suitable precursor for natural rubber [19]. The plant which provides the greatest quantity of latex for commercial utilization is *Hevea brasiliensis,* a tree native to South America which is now found primarily in Malaysia [19]. Natural rubber is a thermoplastic elastomer with desirable mechanical properties of high elongation and good impact strength. Blends of rubber and PLA are immiscible, as PLA is a polar material and rubber is non-polar [20]. Depending on the desired mechanical properties of the final product, immiscibility can either be utilized or overcome. It has been shown that a 10%

NR/PLA blend increases the elongation at break of the virgin polymer by a factor of 40 [15]. Depending on the desired outcome, an ideal percentage of NR in the PLA matrix exists; due to coalescence of rubber particles hindering the positive effects of the rubber phase, increasing the natural rubber percentage to 20% results in a lower elongation at break than a blend with a smaller percentage of NR [15].

PLA and NR blends are immiscible, and the interfacial tension between the two polymers is high. Compatibilizing materials have been shown to reduce interfacial tension between phases in immiscible polymer blends. Graft copolymers are a traditional and effective method to compatibilize [21]. Grafting polar polymers onto NR's isoprene backbone is one method to produce a PLA/NR blend with greater miscibility. Zhang et al. grafted poly(butyl acrylate) onto natural rubber and found that subsequent blends with PLA were miscible, as evidenced by electron microscopy studies. The resulting blends toughened PLA more effectively than blends using virgin NR [20]. Throughout the past 15 years, nanoclays have been extensively researched for the property enhancement that they bring about in polymer/clay nanocomposites as well as the compatibilization effect that they have in polymer blends [22-24]. Because of the unique structure of the montmorillonite class of nanoclays and the organic modifications made to the base clay, intercalation and exfoliation of these nanostructures in the polymer matrix can result in enhanced barrier properties, improved strength, and more. Bitinis et al. studied the effect of adding nanoclays Cloisite CNa^+, C15A, and C30B on the rheological, mechanical, and thermal properties of PLA/NR blends [14]. In this work, we will produce PLA/NR blends with NR percentages of 5, 10, and 15%. The interfacial tension between phases and viscoelastic properties are better understood by utilizing rheological measurements. The final mechanical and thermal properties of the blend will be studied through tensile testing, DSC, and TGA.

Materials and Methods

PLA was provided by NatureWorks. The grade used in this study is 4032D, a film grade. The organoclays Cloisite 30B, 15A, and 10A were from Southern Clay products. All materials were used as received. PLA and clay were dried at 60°C under vacuum for ~12 hours to remove excess moisture. After drying, the materials were placed in a desiccator until use. PLA/NR/MMT blend nanocomposites were prepared in a 200mL capacity counter-rotating mixer operating at 32 rpm. Processing temperature was 190°C and processing time was 5-7 minutes.

For rheological and mechanical testing, samples were prepared using a Carver hydraulic press at 190°C. Thermal analysis was performed on samples obtained directly from the blend.

Characterization

Rheology

Rheological tests were carried out on TA Instruments ARES G2 rheometer with 25mm parallel plate geometry using TRIOS software. A strain sweep was performed on samples to determine the linear viscoelastic region. A frequency sweep was carried out to determine storage and loss moduli of the blends and composites during processing conditions.

Thermal Analysis

Thermogravimetric analysis was performed on TA Instruments SDT Q600. 10-15mg samples were heated from room temperature to 600°C at a rate of 10°C/minute. Differential scanning calorimetry was done on TA Instruments DSC Q2000. Samples were ramped at 10°C/minute to 200°C, cooled, and reheated. The second heating data was analyzed to determine glass transition and melting temperatures. Qseries Advantage software was used to run the TA equipment, as well as to analyze the data.

Mechanical Testing

Tensile testing was performed on Instron 5966 Universal Testing System using Bluehill 3 software. ASTM D638 standard dumbbell samples were cut from 1mm thick sheets. Tensile strength and elongation at both yield and break were determined according to the ASTM standards.

References

[1] M. Ajioka, K. Enomoto, K. Suzuki and A. Yamaguchi, "The basic properties of poly(lactic acid) produced by the direct condensation polymerization of lactic acid," *Journal of Polymers and the Environment,* vol. 3, pp. 225-234, 1995.

[2] D. W. Grijpma, A. J. Nijenhuis, P. G. T. Wijk and A. J. Pennings, "High impact strength as-polymerized PLLA," *Polymer Bulletin,* vol. 29, pp. 571-578, 1992.

[3] M. Jiao, K. Yang, W. Pan, W. Zhang and S. Xie, "Preparation and Characterization of Poly(L-lactic acid)/Polybutadiene Block Copolymers," *Journal of Macromolecular Science, Part B,* vol. 49, pp. 886-896, 2010.

[4] S. C. Schmidt and M. A. Hillmyer, "Synthesis and Characterization of Model Polyisoprene,àíPolylactide Diblock Copolymers," *Macromolecules,* vol. 32, pp. 4794-4801, 1999.

[5] D. Cohn and A. H. Salomon, "Designing biodegradable multiblock PCL/PLA thermoplastic elastomers," *Biomaterials,* vol. 26, pp. 2297, 2005.

[6] L. V. Labrecque, R. A. Kumar, V. Davé, R. A. Gross and S. P. McCarthy, "Citrate esters as plasticizers for poly(lactic acid)," *J Appl Polym Sci,* vol. 66, pp. 1507-1513, 1997.

[7] N. Ljungberg and B. Wesslén, "Tributyl citrate oligomers as plasticizers for poly (lactic acid): thermo-mechanical film properties and aging," *Polymer,* vol. 44, pp. 7679-7688, 2003.

[8] M. Baiardo, G. Frisoni, M. Scandola, M. Rimelen, D. Lips, K. Ruffieux and E. Wintermantel, "Thermal and mechanical properties of plasticized poly(L-lactic acid)," *J Appl Polym Sci,* vol. 90, pp. 1731-1738, 2003.

[9] Z. Kulinski, E. Piorkowska, K. Gadzinowska and M. Stasiak, "Plasticization of Poly(l-lactide) with Poly(propylene glycol)," *Biomacromolecules,* vol. 7, pp. 2128-2135, 2006.

[10] M. E. Broz, D. L. VanderHart and N. R. Washburn, "Structure and mechanical properties of poly(d,l-lactic acid)/poly(Îμ-caprolactone) blends," *Biomaterials,* vol. 24, pp. 4181, 2003.

[11] I. Noda, M. M. Satkowski, A. E. Dowrey and C. Marcott, "Polymer Alloys of Nodax Copolymers and Poly(lactic acid)," *Macromolecular Bioscience,* vol. 4, pp. 269-275, 2004.

[12] G. Chen, H. Kim, E. Kim and J. Yoon, "Compatibilization-like effect of reactive organoclay on the poly(l-lactide)/poly(butylene succinate) blends," *Polymer,* vol. 46, pp. 11829, 2005.

[13] L. Jiang, M. P. Wolcott and J. Zhang, "Study of Biodegradable Polylactide/Poly(butylene adipate-co-terephthalate) Blends," *Biomacromolecules,* vol. 7, pp. 199-207, 2006.

[14] N. Bitinis, R. Verdejo, E. M. Maya, E. Espuche, P. Cassagnau and M. A. Lopez-Manchado, "Physicochemical properties of organoclay filled polylactic acid/natural rubber blend bionanocomposites," *Composites Sci. Technol.,* vol. 72, pp. 305, 2012.

[15] N. Bitinis, R. Verdejo, P. Cassagnau and M. A. Lopez-Manchado, "Structure and properties of polylactide/natural rubber blends," *Mater. Chem. Phys.,* vol. 129, pp. 823, 2011.

[16] S. Ishida, R. Nagasaki, K. Chino, T. Dong and Y. Inoue, "Toughening of poly(L-lactide) by melt blending with rubbers," *J Appl Polym Sci,* vol. 113, pp. 558-566, 2009.

[17] R. Jaratrotkamjorn, C. Khaokong and V. Tanrattanakul, "Toughness enhancement of poly(lactic acid) by melt blending with natural rubber," *J Appl Polym Sci,* vol. 124, pp. 5027-5036, 2012.

[18] T. M. Lewinsohn, "The geographical distribution of plant latex," *Chemoecology,* vol. 2, pp. 64-68, 1991.

[19] H. Mooibroek and K. Cornish, "Alternative sources of natural rubber," *Appl. Microbiol. Biotechnol.,* vol. 53, pp. 355-365, 2000.

[20] C. Zhang, C. Man, Y. Pan, W. Wang, L. Jiang and Y. Dan, "Toughening of polylactide with natural rubber grafted with poly(butyl acrylate)," *Polym. Int.,* vol. 60, pp. 1548-1555, 2011.

[21] C. Koning, M. Van Duin, C. Pagnoulle and R. Jerome, "Strategies for compatibilization of polymer blends," *Progress in Polymer Science,* vol. 23, pp. 707-757, 1998.

[22] M. Si, T. Araki, H. Ade, A. L. D. Kilcoyne, R. Fisher, J. C. Sokolov and M. H. Rafailovich, "Compatibilizing Bulk Polymer Blends by Using Organoclays," *Macromolecules,* vol. 39, pp. 4793-4801, 2006.

[23] S. Vandebril, J. Vermant and P. Moldenaers, "Efficiently suppressing coalescence in polymer blends using nanoparticles: role of interfacial rheology," *Soft Matter,* vol. 6, pp. 3353-3362, 2010.

[24] S. Sinha Ray and M. Bousmina, "Effect of Organic Modification on the Compatibilization Efficiency of Clay in an Immiscible Polymer Blend," *Macromolecular Rapid Communications,* vol. 26, pp. 1639-1646, 2005.

(315b) Bioplastics and Sustainability

Amar K. Mohanty, Department of Plant Agriculture & School of Engineering, University of Guelph, Guelph, ON, Canada

Biobased or green economy is challenging to agriculture, forestry, academia, government and industry. The incorporation of bio-resources, e.g. crop-derived green plastics and plant derived natural fibres into composite materials are gaining prime importance in designing and engineering green composites. Biocomposites derived from natural fibers and traditional polymers like polypropylene, polyethylene, epoxy and polyesters have been developed for automotive parts and building structures. Renewable resource based bioplastics like polylactic acid (PLA), polyhydroxyalkanoates (PHAs), biobased polytrimethylene terephthalate (PTT), bio-polyolefin, bio-polyamide, cellulosic plastics, soy protein based bioplastics and vegetable oils derived bioresins need value-added and diverse applications to compete with the fossil fuel derived plastics. Through reactive blends, composites and nanocomposites new biobased materials are under constant development.

Natural fibres are lighter, less expensive, have superior specific strength, require comparatively less energy to produce, are good for the environment, biodegradable and have superior sound abatement characteristics as compared to synthetic glass fibres. Hybrid and intelligently engineered green composites are going to be the major drivers for sustainable developments of new industrial bioproducts. Besides agricultural natural fibers like kenaf, jute, flax, industrial hemp, and sisal; inexpensive biomasses such as wheat straw, rice stalks, corn stovers, grasses, soy stalks and lignin have great potential for use in sustainable biobased materials.

This presentation will highlight the current status, opportunities and challenges of bioplastics and biobased materials for uses in car parts, consumer goods and sustainable packaging.

Acknowledgements: Ontario Ministry of Agriculture, Food and Rural Affairs (OMAFRA)-New Directions Research Programs; University of Guelph-OMAFRA (Bioeconomy for Industrial Uses); Natural Sciences and Engineering Research Council of Canada (NSERC)-Discovery Grant; and Auto21.

POLYBUTYLENE SUCCINATE LIFE CYCLE ASSESSMENT VARIATIONS AND VARIABLES

Hassan I. Moussa, ERS, University of Waterloo, Waterloo, ON, Canada,
Steven B. Young, SEED, University of Waterloo, Waterloo, ON, Canada
Yves Gerand, Information and Technology for Agricultural Processes, Irstea, Montpellier, France

Introduction

Polybutylene succinate (PBS) is a novel aliphatic polyester polymer that was originally developed for its biodegradability, physical and mechanical properties, thermal and chemical resistance and processability. PBS is versatile, stable [1,2,3] and purported to be totally biodegradable under normal conditions [4]. PBS approximates common thermoplastics like polypropylene in its performance and therefore presents tremendous possibilities.

It is essential to acknowledge from the outset that bio-based materials may not be in many aspects environmentally preferable to their conventional counterparts; moreover, it should be emphasised that that claims of "overall" superiority or "sustainability" are inappropriate and cannot be supported by science. Rather an indicator-by-indicator comparison should be considered, with attention to the uncertainty and variably of data and therefore to the conclusiveness of any results.

In 1990 ICI commercialized the first synthesized biodegradable plastic, "Biopol" [5]. Biodegradable plastics were assumed to be introduced in order to reduce the environmental burdens associated with concerns over the landfill volume that plastics occupy [6]. In recent years, bio-degradable plastics and plastics made from bio-based resources have been introduced such as poly butylene succinate (PBS) and polylactic acid (PLA) [6].

While not all bio-based polymers are bio-degradable and not all bio-degradable polymers are bio-based, both are confusingly called bio-plastics [7]. According to Tokiwa et al., (2009) there are three types of bio-plastics: bio-based and non-biodegradable such as polyethylene synthetized from bio-mass or renewable sources; bio-based and bio-degradable such as poly lactic acid (PLA) [9], and poly butylene succinate PBS; and non-bio-based and bio-degradable such as poly butylene succinate PBS, and polycaprolactone (PCL) [8]. This study concentrates only on PBS from type 3, the non-biobased, and bio-degradable.

The synthesized bio-degradable polymers can be categorized in two groups based on physical, chemical and structural classification. The first group is similar to the traditional light density polyethylene (PE-LD), being highly crystallized and flexible. The second group is similar to polypropylene (PP), polystyrene (PS) and polyethylene terephthalate (PET), as it is semi-amorphous and more rigid [16]. Poly (butylene succinate) (PBS), the subject of our study, falls under the second group of bio-degradable polymers.

One of the challenges for conventional plastics and their products (e.g., polypropylene, polyethylene) is at their end-of-life when they are recycled, incinerated or treated as solid waste in landfills. While conventional plastics can be recycled and re-used, there are restrictions when it comes to reusing them for food packaging [10]. Health Canada published guidelines addressing concerns for using recycled plastics in food packaging due to the permeablity of plastics in "Division 23 of the Food and Drugs Regulation" [11]. Incineration is another approach but there are environmental concerns of emissions from incinerators. Furthermore, concerns regarding the decreasing availability of landfills around the world were also expressed [10]. Most importantly, however, is the fact that conventional plastics do not degrade easily in natural or composting landfill environments. Consequently, conventional plastics are considered a threat to the environment [8,12].

In recent years, global chemical manufacturers have begun developing and commercializing a wide variety of bio-degradable synthetic polyesters including PBS and PLA claiming that they may provide some solutions for the end-of-life environmental management of plastics [8,13,14]. In addition, bio-degradable polymers may provide solutions for health problems associated with the recycling and incineration of conventional plastics waste [10,11]. A depolymerization process is the degradation of bio-degradable polymers that results with low weight molecular monomers, dimers and oligomers. The depolymerisation process occurs under the effect of micro-organisms enzymes, biologically, or photodegradable / oxo-degradable [15] and produces carbon dioxide under aerobic conditions or methane under anaerobic conditions [7,8,10].

A study by Malveda, Löchner, and Yokose (2010) on bio-degradable polymers indicated that the annual consumption growth of bio-degradable polymers in North America, Europe and Asia is expected to reach 13% in the year 2014. This annual growth is influenced mainly by increasing demand from the packaging industry. In North America, biodegradable polymer consumption growth is essentially driven by competitive pricing, continuous improvement in processability, overall competitive properties of bio-degradable polymers and concerns due to the end-of-life of non-bio-degradable polymers on local, state and federal levels [17].

While a modest body of literature has accumulated over the past decade on the mechanical and physical properties and the bio-degradability of PBS and its composites [3,8,18,19,20,21], few studies have addressed the potential environmental impacts throughout the life-cycle of the material [4,22]. Therefore, this study considers the potential environmental impacts associated with the manufacturing of 100% petroleum-based PBS.

Method

Life cycle assessment (LCA) was used to characterize the environmental profile of this polymer as part of a broader project on sustainability characterization of biobased materials. The production of succinic acid includes three unit processes: production of maleic anhydride, succinic anhydride and succinic acid. The following possible variations in feedstock and energy sources were considered:

1. 100% fossil sources used for both the succinic acid and 1,4-butanediol (1,4 BDO). Succinic acid is produced via the hydrogenation of maleic anhydride followed by the hydrolysis of succinic anhydride. Maleic anhydride can be produced either from the direct oxidation of benzene or catalytic oxidation of n-butane. 1,4 BDO is produced using acetylene and formaldehyde.
2. The polymerization of succinic acid and 1,4 BDO produces the polymer, 100% bio-based feedstock that is converted into succinic acid via fermentation of sugar. Commercial producers include Myriant, DSM, and BioAmber. 1,4 BDO can also be derived via the hydrogenolysis process.
3. 1,4 BDO can be produced directly from sugar or biomass, as produced by Genomatica,

and hybrid combinations of bio-based succinic acid and fossil-based 1,4 BDO. The hybrid system is an option since technology for manufacturing bio-based 1,4 BDO is still in its early stages and no manufacturer had begun producing it for commercial purposes.

Impact Assessment Indicator Methods

Consistent with current practices in LCA, several the life-cycle impact assessment methods employed in the study. Cumulative Energy Demand (CED) represents an inventory-based indicator. The global warming potential (GWP) for climate change forcing is based on Intergovernmental Panel on Climate Change (IPCC) characterisation factors for direct greenhouse gas propensity of emissions over 100 years. The Tool for the Reduction and Assessment of Chemical and Other Environmental Impacts (TRACI) from the US EPA is a life cycle impact assessment method that characterizes a number of important impact categories using nine impact indicators that are calculated at the "midpoint level" of environmental interaction. In addition to GWP, this study shows results for the indicators of acidification ("acid rain"), non-carcinogenic toxicity and eco-toxicity, which are each represented on an equivalency basis [23,24].

Data and Calculations

The study employed SimaPro software for the LCA and relied on background data from the ecoinvent database. Challenges due to data quality and data availability for such new processes and feedstock create uncertainties in the analysis of potential environmental profiles. The lack of primary and empirical data related to the manufacturing of PBS, necessitated the use of secondary and tertiary data to characterize environmental impacts of the PBS manufacturing process. This study relied on PBS data from the polymer "Bionolle" and the respective model for its production: Bionolle is a 100% petroleum-based PBS manufactured by the major manufacturer of this polymer, Showa Denko of Japan [1,25]. PBS is used in manufacturing of agricultural, packaging, and civil engineering materials and was recently tested for manufacturing of automotive parts [22].

Limitations

The life-cycle was characterised from source of resources to the production of plastics pellet (cradle-to-gate), and did not consider the effects of further processing, use, or end-of-life. Thus, biodegradability is not covered in this analysis. The precision of numerical results are constrained to two significant figures. A sensitivity analysis was undertaken that focused on the variables of insulation efficiency and polymerization yield for hydrogenation and polymerization reactors. The polymerization yields used in the sensitivity analyses were 99%, 95%, and 85%; whereas the insulation efficiencies used were 95%, 60% and 0%.

Study results

Cumulative Energy Demand (CED) Indicator

Results indicated a CED of 140 MJ/kg PBS pellet across the process-level contributors (see Figure 1). Cumulatively, the three unit processes that represent the production of succinic acid (including maleic anhydride and succinic anhydride) have a bigger value than that for BDO.

Global Warming Potential (GWP) Indicator

The study revealed that the total GHG emission resulting from the production of 1 kg PBS dry pellets is 6.6 kg CO2-eq as distributed across the main processes (Figure 2). GHG emissions compare to a previous analysis by Showa Denko K.K. that assessed a similar but different material with varying assumptions on thermal and electrical energy, biomass feedstock cultivation, and system boundaries.

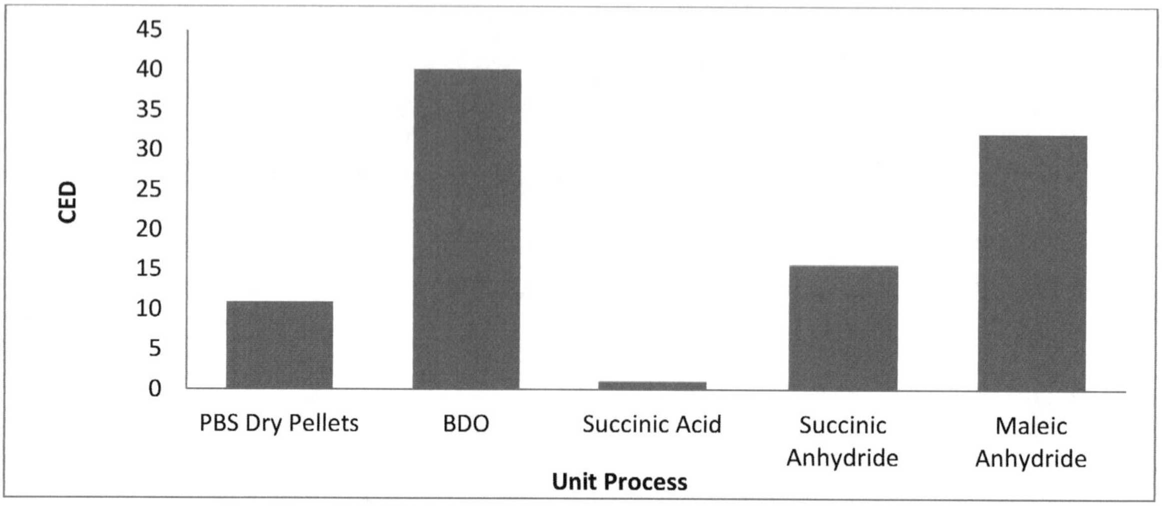

Figure 1: CED indicator contributions (of 140 MJ) for the production of 1 kg PBS dry pellets.

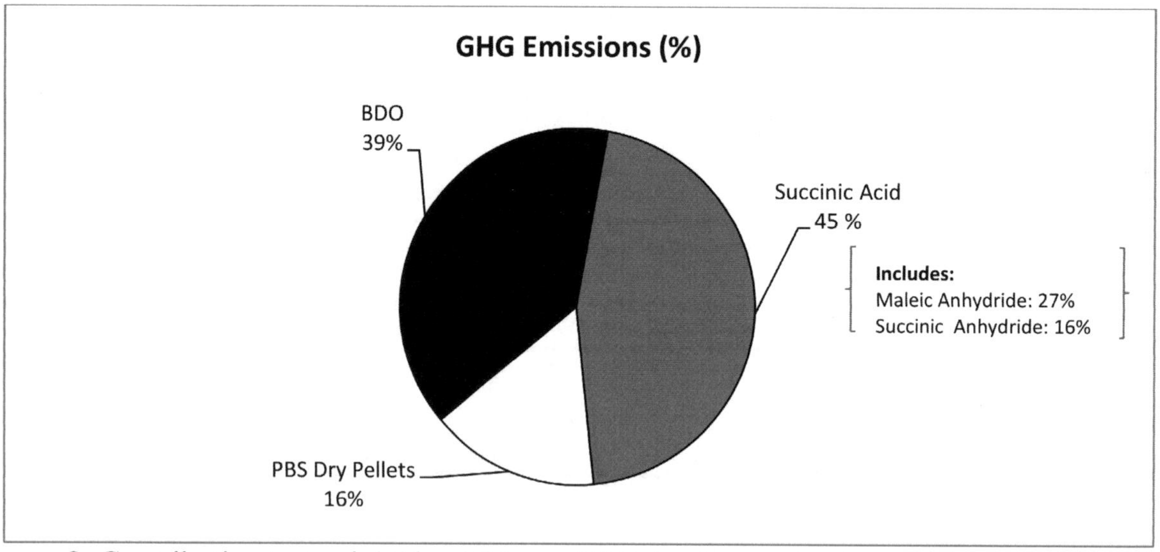

Figure 2: Contribution to total 6.6 kg CO2-eq. GWP for the production of 1 kg PBS dry pellets.

TRACI Indicators

Results for TRACI are shown in Table 1 for the production of 1 kg PBS dry pellets. The other TRACI indicators are negligible, for their values are approximately zero. The contribution of unit processes to the total acidification and eco-toxicity is shown in Figure 3. The production of succinic acid contributes most substantially, followed by the BDO process. The contribution of unit processes to the total non-carcinogenic toxicity indicator shows that succinic acid is the largest contributor (Figure 4).

Table 1: Characterization results of TRACI indictors for 1 kg PBS dry pellets.

Impact indicator	Unit	Total	PBS Dry Pellets	PBS Wet Pellets	PBS 3.6% Water	BDO	Succinic Acid
Acidification potential	H+ moles-eq.	1.4	0.0	0.1	0.1	0.4	0.8
Non-carcinogenic toxicity potential	kg toluene-eq.	160	0.1	45	15	36	63
Eco-toxicity potential	kg 2,4-D eq.	11	0.0	0.9	1.1	4.1	4.5

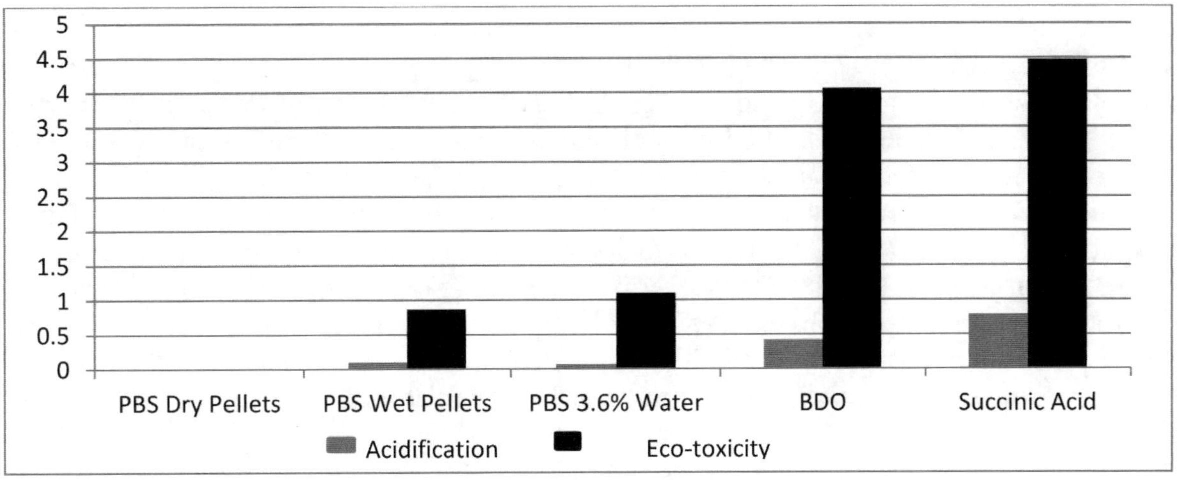

Figure 3: Relative unit processes contributions to the total acidification and eco-toxicity indicators for 1 kg PBS dry pellets

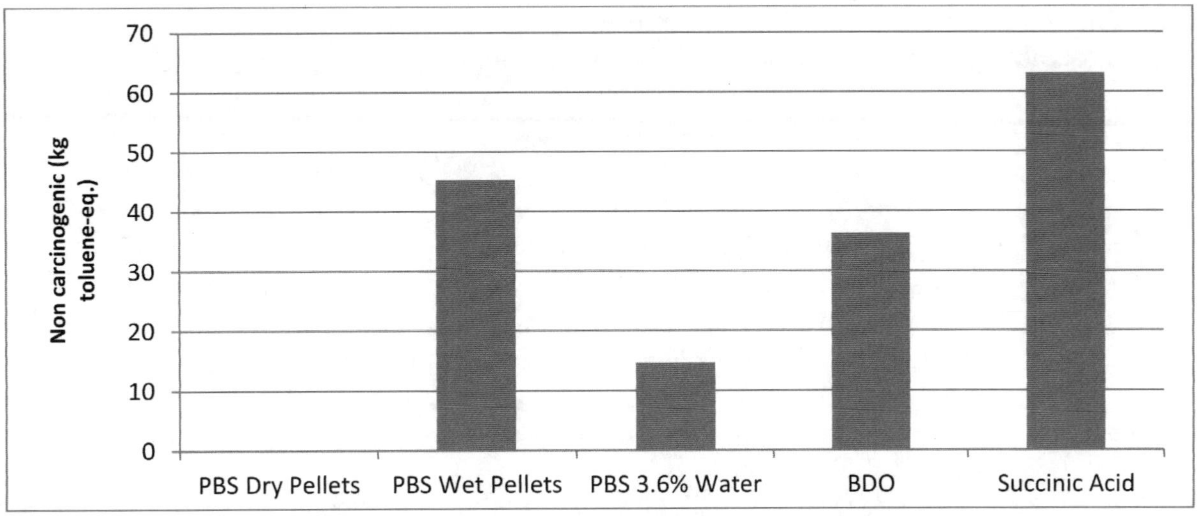

Figure 4: Unit processes contribution to the total non-carcinogenic toxicity materials as a result for the production of 1 kg PBS dry pellets.

Discussion

Questions of feedstocks and energy dominate considerations in the LCA results. Sensitivity analyses showed that variations in process yields at the hydrogenation and polymerization reactors resulted in minor linear responses in the LCA indicator results. More import than feedstock efficiencies, however, was energy utilisation in the process. The efficient utilization of waste energy, for example, through heat recovery or reactor insulation, can drive better indicator results. This is somewhat intuitive: energy use is reflected directly in the CED indicator, and correlates strongly to the greenhouse gases associated with the GWP indicator, and less-so to the air and multi-media environmental quality indicators (acidification and toxicity). Energy use has a strong influence on the environmental LCA results.

The desire to move from fossil-based to biobased materials often presumes an environmental preference associated with the latter. Polymer feedstocks may be derived from both fossil-based and biobased sources. Energy used in chemical processes, similarly, may be obtained from both fossil-based and biobased sources. The current study suggests that process energy use is a dominant driver for LCA indicator results. Therefore it is valid to question whether feedstock type or energy type will be more important in the fuller environmental sustainability analysis. Further work, it is suggested, should be designed to separate the variables of both feedstock/energy, and fossil-based/biobased – and perhaps also to consider hybrid fossil-biobased systems. Variations along these lines, supported by rigorous LCA studies using high quality data will better serve evidence-based decisions, rather than preconceived assumptions, regarding the sustainability performance of biobased systems.

Acknowledgements

This research was supported by the Province of Ontario, Ontario Ministry of Agri-foods and Rural Affairs (OMAFRA), Project Number: 200004-200005-200006, Renewable, Recyclable and Lightweight Hybrid Green Composites from Lignin, Switch grass, Miscanthus and Bioplastics for Sustainable Manufacturing.

References

1. Showa Denko K.K. (retrived 2012, March 29). *Bionolle*. Retrieved from Showa Denko K. K.: http://www.sdk.co.jp/english/products/119/125/12907.html
2. Tserki, V., Matzinos, P., Pavalidou, E., Vachliotis, D., & Panayiotou, C. (2006). Biodegradable aliphatic polyesters. Part I. Properties and biodegradation of poly(butylene succinate--butylene adipate). *Polymer Degradation and Stability, 91*(2), 367-376.
3. Kim, M.-N., Kim, K.-H., Jin, H.-J., Park, J.-K., & Yoon, J.-S. (2001). Biodegradability of ethyl and n-octyl branched poly(ethylene adipate) and poly(butylene succinate). *European Polymer Journal, 37*(9), 1843-1847.
4. Ichikawa, Y., & Mizukoshi, T. (2011). Bionolle (Polybutylenesuccinate). *Advanced Polymer Science Journal, 245*, 285-314. doi:10.1007/12_2011_145

5. British Plastics Federation [BPF]. (retrieved 2012, April 03). *A History of Plastics*. Retrieved from British Plastics Federation - Plastipedia: http://www.bpf.co.uk/plastipedia/plastics_history/default.aspx
6. Gironi, F., & Piemonte, V. (2011). Life Cycle Assessment of Polylactic Acid and Polyethylene Terephthalate Bottles for Drinking Water. *Environmental Progress & Sustainable Energy, 30*(3), 459-468. doi:10.1002/ep.10490
7. Chin, I.-J., & Uematsu, S. (2011). Biodegradation of Polymeric Systems. In V. Mittal, *Nanocomposites with Biodegradable Polymers* (pp. 28-57). New York, U.S.: Oxford University Press.
8. Tokiwa, Y., Calabia, B., Ugwu, C., & Aiba, S. (2009). Biodegradability of Plastics. *International Journal of Molecular Sciences, 10*(9), 3722-3742. doi:10.3390/ijms10093722
9. Corn Refiners Association. (retrieved 2012, April 12). *Bioproducts*. Retrieved from Corn Refiners Association: http://www.corn.org/products/bioproducts/
10. Bohlmann, G. (2005). General Characteristics, Processability, Industrial Application and Market Evolution of Biodegradable Polymers. In C. Bastioli, *Handbook of Biodegradable Polymers* (pp. 183-218). Shawbury: Rapra Technology Limited.
11. Stewardship Ontario. (2005). *Ontario Blue Box Market Overview: 2002/2003 Blue Box Materials Generated and Municipally Marketed - Mass Balance Report.* Toronto, Canada: Stewardship Ontario.
12. Huda, M., Drzal, L., Mohanty, A., & Misra, M. (2008). Effect of fiber surface-treatment on the properties of laminated biocomposites form poly(lactic acid) (PLA) and kenaf fibers. *Composites Science and Technology, 68*(2), 424-432.
13. Kim, H.-S., Yang, H.-S., & Kim, H.-J. (2005). Biodegradability and Mechanical Properties of Agro-Flour–Filled Polybutylene Succinate Biocomposites. *Journal of Applied Polymer Science, 97*(4), 1513-1521.
14. Zheng, Y., Yanful, E.-K., & Bassi, A.-S. (2005). A Review of Plastic Waste Biodegradation. *Critical Reviews In Biotechnology, 25*(4), 243-250.
15. Gautam, S.-P. (2009). *Bio-degradable Plastics- Impact on Environment.* Delhi, India: Central Pollution Control Board - Ministry of Environemnt and Foresets.
16. Leaversuch, R. (2002, September). *Biodegradable Polyesters: Packaging Goes Green.* (R. Leaversuch, Ed.) Retrieved from The Online Magazine of Plastics Technology: http://www.ptonline.com/articles/biodegradable-polyesters-packaging-goes-green
17. Malveda, M., Lochner, U., & Yokose, K. (2010, January). *Biodegradable Polymers.* Englewood, USA.: IHS Chemicals. Retrieved from http://chemical.ihs.com/CEH/Public/Reports/580.0280
18. Wan, C., & Chen, B. (2012). Reinforcement of biodegradable poly(butylene succinate) with low loadings of graphene oxide. *Journal of Applied Polymer Science.* doi:DOI: 10.1002/app.38136
19. Zhu, W., Wang, X., Chen, X., & Xu, K. (2010). Miscibility, Crystallization, and Mechanical Properties of Poly(3-hydroxybutyrate-co-4-hydroxybutyrate)/Poly(butylene succinate) Blends. *Journal of Applied Polymer Science, 117*(2), 838-848.
20. Zhao, J., Wang, X., Zeng, J., Yang, G., Shi, F., & Yan, Q. (2005). Biodegradation of poly(butylene succinate--butylene adipate). *Polymer Degradation and Stability, 90*(1), 173-179.
21. Fujimaki, T. (1998). Processability and properties of aliphatic polyesters, 'BIONOLLE', synthesized by polycondensation reaction. *Polymer Degradation and Stability, 59*, 209-214.

22. Terasawa, I., Tsuneoka, K., Tamura, A., & Tanase, M. (2008). Development of Plant-Based Plastic Technology, 'Green Plastic'. *Mitsubishi Motors Technical Review*(20), pp. 91-96.
23. Bare, J. (2011). TRACI 2.0: the tool for the reduction and assessment of chemical and other enviornmental impacts 2.0. *Clean Technologies and Environmental Policy, 13*(5), 687-696.
24. PRé Consultants. (2008). *SimaPro 7: Database Manual - Methods Library.* the Netherlands: PRé Consultants.
25. Jun, X., & Guo, B.-H. (2010). Plastics from Bacteria: Natural Functions and Applications - Microbial Succinic Acid , Its Polymer Poly (butylene succinate), and Applications. In A. Steinbuchel (Ed.), *Microbiology Monographs* (Vol. 14, pp. 347-388). Berlin: Springer.

(315d) Creep and Aging of Phbv-Natural Fiber Composites

Sabbie Miller, Dept. of Civil & Environmental Engineering,, Stanford University, Stanford and Sarah Billington, Dept. of Civil & Environmental Engineering, Stanford University, Stanford, CA

A biobased composite composed of poly(β-hydroxybutyrate)-co-(β-hydroxyvalerate) (PHBV) matrix and natural reinforcing fibers has a potential closed-loop life cycle as their constituents can potentially produce their own feedstock carbon upon anaerobic degradation to produce more polymer. These closed loop materials have the potential to replace conventional polymer because their mechanical properties are comparable to engineered woods and several types of glass fiber reinforced polymers However their long-term, in-service performance is largely unknown. As a semicrystalline polymer, PHBV is susceptible both to viscoelastic deformation under sustained loading aging effects. In this research, the viscoelastic behavior of PHBV/natural fiber composites is being analyzed through a series of creep and creep-recovery experiments. Isochronous deformation of the PHBV composites is compared to that of similar materials. In addition, the effect of polymer aging on the tensile properties of PHBV/natural fiber composites is under investigation. Both experimental and modeling results will be presented.

(315e) Synthesis of Sustainable and Stable Latex of Biopolymer Encapsulated Nanolignin-Clay Hybrid Via Miniemulsion Polymerization

Zhaohui Tong, Agricultural and Biological Engineering Department, IFAS,, University of Florida, Gainesville, FL and Suguna Jairam, University of Florida, Gainesville, FL

Synthesis of sustainable and stable latex of biopolymer encapsulated nanolignin-clay hybrid via miniemulsion polymerization

Suguna Jairam, Zhaohui Tong*

Agricultural and Biological Engineering Department, IFAS, University of Florida,

Gainesville, FL, 32611-0570

Email: ztong@ufl.edu

Everyday usage of water-borne dispersions (latex) cannot be understated, with wide areas of application ranging from usage in packaging, medicine, to food and structure. Stable and sustainable latex from renewable resources will attract a lot interest due to its advantages of better environmental concern, easy manipulation and energy saving. In addition, the lack in enhanced functionality of these latexes demands an improvement of their properties such as anti-microbial properties, barrier to heat, gas and moisture, UV-protection and mechanical properties. Therefore, this project aims to synthesize a novel water-borne dispersion encapsulating nano-lignin/clay hybrid in a bio-based acrylated polymer derived from bio-based glycerol and/or free fatty acids from waste stream of biodiesel. Preliminary studies were performed on the synthesis of nanolignin-clay hybrid and the encapsulation of this hybrid into polymer to form stable latex. The lignin extracted from bioethanol process was modified to quaternary ammonium lignin and it can be loaded on the clay surface to produce an organophilic hybrid. FTIR analysis of clay before and after lignin addition showed successful binding and thermogravimetric analysis of the hybrid showed about 32% (by weight) loading of lignin on clay. Furthermore, miniemulsion polymerization was applied for the encapsulation of organophilic hybrid into both petroleum- and bio-based polymer and stable latex was formed. X-ray diffraction demonstrated the intercalation of nanohybrid in the polymer matrix. The property results showed that the encapsulation of nanolignin-clay hybrid improved tensile strength and thermal stability of final latex film.

(315f) Innovative Bio-Resin Compounding of Locally Grown Ag Fibers with Post-Consumer Plastic

Atul Bali, CGTech, Brampton, ON, Canada

CGTech has earned an exclusive license from the University of Guelph in this regard for specific bio-resin formulations.. These bio-resins are a first of their kind in the world and have already gained market acceptance as demonstrated by being on the shelves of Home Hardware stores as bio-composite storage bio-bins under the NATURA brand of the customer.

CGTech has market access in both auto and consumer market segments. VW in the auto sector and The Home Depot, Rona and Home Hardware in the consumer products area are market leaders. CGTech has been engaged by all these companies to develop and market polymer formulations using waste streams of plastic (post-consumer) and Agricultural fibres. For example, working closely with the injection molder, the Uof Guelph, and Home Hardware, bio-resin was compounded and provided to Embee Plastics, Linwood, ON for supply of five different sizes of bio-bins (storage bins) to Home Hardware, as cited above - meeting the price-performance criteria.

The market potential is huge for this first of its kind technology as can be well –contemplated by anyone who is familiar with this business of the transition to competitive biomaterials to realize a sustainable green future. CGTech is well-poised to leverage this opportunity.

(315g) A Fresh LOOK At the Potential of Indian Lantana Plants for Its Profitable Utilization

Amiya Kumar, Department of Paper Technology, Indian Institute of Technology Roorkee, Roorkee,India, Saharanpur, India

A FRESH LOOK AT THE POTENTIAL OF INDIAN LANTANA PLANTS FOR ITS PROFITABLE UTILIZATION

Amiya.K. Ray,Vivek Kumar,Manish Kumar Puri and Vikas Kumar Agarwal

Indian Institute of Technology Roorkee,

Saharanpur Campus,India

ABSTRACT

One of the most stringent problems in Indian paper mill is the acute scarcity of virgin fibres from pulpable raw materials along with other constraints.Therefore there are alround efforts to combat this problem by intensive search for exploring new renewable raw materials of right quality available indigenously to make pulp. Lantana, a shrub, perennial in nature, is available in large quantity in almost every part of this country. However, there are varieties of this species *of Verbenaceae/ Vervain* family identified in India such as Lantana cacamara,Lantana aculeta,Camara vulgaris, Lantana scabrida , Lantana scandes etc. Earlier some experimental and kinetic studies are made with the variety –Lantana Camara. However,more extensive studies are required to evaluate its potential without going in to the identification of species. A large number of lantana plant from three different locations,from Western UP and Uttarakhand in India are collected and randomly blended. The blended raw materials are found to possesss an average value of hollocellulose, 74.3%, extractives,9.6%, lignin, 18.3% , Pentosans ,16.8 % and silica,2.45% which indicated that it can be a good substitute for bamboo or hardwood for paper making. Furher, the leaves of this plant can be good medicinal source. Extensive experimental trials for pulping are carried out in the laboratory based on statistical full factorial design followed by kinetic studies. Statistical multivariable regression equations in terms of screened yield, kappa number, and lignin are developed for design of the process and for further simulation. As H factor is the most important control parameter in pulping process which in turn depends on both time and temperature schedule , the detailed kinetic studies with these lantana of arbitrary species are carried out at the same conditions of the earlier experiments on Lantana camera. The specific reaction rate, activation energy, reaction order, H factor and Tau factor are evaluated and Kinetic rate equations are established. The proposed models are then compared with the previously predicted models. Multivariable regression equations also are developed to interrelate the process and design parameters and the results were compared.

(477a) Techno-Economic Evaluation of Integrated Renewable Acetic Acid, Glycolaldehyde, and Acetol Production From Biomass-Derived Pyrolysis Oil

André B. de Haan, Caecilia R. Vitasari and G. Wytze Meindersma, Chemical Engineering, Eindhoven University of Technology, Eindhoven, Netherlands

Acetic acid, glycolaldehyde, and acetol are bio-based chemicals which can be isolated from pyrolysis oil in an integrated biorefinery. This paper discusses the conceptual design of an integrated separation process for forest residue- and pine-derived pyrolysis oils. The process simulation was performed in Aspen Plus®, while the equipment cost was estimated in Aspen Process Economic Analyzer. The process was designed for a capacity of 200 kton pyrolysis oil per year and involves extraction, distillation, and evaporation. Water and 2-ethyl-1-hexanol are used as extraction solvents. All process step designs are based on data derived from experiments with forest residue derived pyrolysis oil. The designed process can isolate more than 99% glycolaldehyde and acetic acid and about two-thirds of acetol present in the oils. In comparison with the forest residue-based process (21 M€), the pine-based process requires a higher capital investment of 23 M€ and a slightly higher production cost 49 M€/a versus 48 M€/a, but is able to provide a higher revenue of 62 M€/a instead of 50 M€/a because pine-derived pyrolysis oil contains more acetic acid, glycolaldehyde, and acetol, which also makes it less sensitive to market price. The plant profitability increases with production capacity up to 300 kton pyrolysis oil per year. In this capacity, pine-derived pyrolysis oil yields approximately a three times higher return-on-investment (ROI) than forest residue-derived pyrolysis oil.

(477b) Clean Pyrolytic Sugars Solution

Marjorie Rover[1], Patrick Johnston[1], Laura Jarboe[2] and Robert C. Brown[3], (1)Center for Sustainable Environmental Technologies, Iowa State University, Ames, IA, (2)Department of Chemical and Biological Engineering, Iowa State University, Ames, IA, (3)Mechanical Engineering, Iowa State University, Ames, IA

This study focuses on the effective removal of contaminants from pyrolytic sugar to produce a suitable fermentation substrate. Pyrolytic sugars contain small percentages of other water soluble constituents such as low molecular weight acids, furfural, phenols, as well as other contaminants. A novel approach to "cleaning" the pyrolytic sugars prior to microbial fermentation has been developed and we will present our recent findings, innovative advances and strategies.

Bio-oil may be the most economical approach to advanced biofuel production. Produced from the fast pyrolysis of biomass, bio-oil contains hundreds of chemical compounds. Among the most commercially important components of bio-oil are sugars and phenols.

Significant progress has been achieved in the advancement of bio-oil collection, allowing for an effective bio-oil fractionation method for the separation of the pyrolytic sugars and the lignin derived constituents. A simple "water wash" removes the soluble sugars from the water insoluble components, providing a potential aqueous fermentable sugar stream.

This research focuses on various strategies and separation techniques to remove contaminants from the pyrolytic sugar stream.

(477c) 5-HMF Production in Piperylene Sulfone: A Bifunctional Solvent for Reaction and Separation

Christopher J Butch[1], Jakob Crowe[1], Gregory Marus[1], Pamela Pollet[2], Charles L. Liotta[2] and Charles A. Eckert[1], (1)School of Chemical and Biomolecular Engineering, Georgia Institute of Technology, Atlanta, GA, (2)School of Chemistry and Biochemistry, Georgia Institute of Technology, Atlanta, GA

5-Hydroxylmethylfurfural has garnered wide interest as a potential bio-based commodity chemical due to its multiple reactive groups and relatively easy formation from fructose and other sugars. For example HMF can be produced from fructose in DMSO in high yields; however the separation is difficult and wasteful. The literature contains numerous examples of HMF production in DMSO or DMSO and water, separated by extraction into methyl isobutyl ketone. When combined with heterogeneous catalysis from ion exchange resins, the yields become particularly attractive. However in each of these cases, despite the good yields, recovery of the product requires expensive liquid-liquid extraction processes, and often the DMSO is not recyclable, thus rendering the reaction impractical.

To address the difficulties associated with separating HMF from DMSO, we use piperylene sulfone, a solvent with very similar solvatochromic properties to DMSO. These similarities make piperylene sulfone a viable replacement as reaction medium. Piperylene sulfone is formed through a reverse cheletropic reaction between sulfur dioxide and 1,3-trans-piperylene. Superiorly to DMSO, piperylene sulfone can be decomposed to sulfur dioxide and piperylene at temperatures of ~110C which allows for facile separation of reaction products. The off-gassed sulfur dioxide and trans-piperylene can be re-reacted elsewhere to reform the piperylene sulfone, enabling recyclability.

We demonstrate a combined reaction separation system composed of fructose in piperylene sulfone with Amberlyst-15 as catalyst. We will present the results of a statistical design of experiments conducted to optimize conversion of HMF as well as the separation of HMF from piperylene sulfone. We shall also present opportunities for future research using this solvent system.

ENHANCED COMPOST DRYING PROCESS OF SEWAGE SLUDGE WITH CARBONIZED SLUDGE

***N. Kobyashi*[1], *Y. Itaya*[2], *L. Li*[2], *A. Suami*[2], *M. Sawai*[3], *H. Hamabe*[3]**

[1]*Department of Chemical Engineering, Nagoya University, Nagoya, Japan*

[2]*Department of Mechanical and Systems Engineering, Gifu University, Gifu, Japan*

[3]*Obu Unity Co. Ltd., Obu, Aichi, Japan*

Introduction

Sewage disposal is an essential environmental conservation technology in civil society but a municipal treatment process consuming a large amount of energy (Slim et al., 2008). In order to mitigate carbon dioxide emission and reduce energy consumption, excellent energy efficient process should be established including recycle of sewage sludge. Sludge as residue contains moisture as high as 80 percent level and is hard to dry. So heating value of sludge is too low to burn out it without using auxiliary fuel in incineration process.

In recent year, there are attempts drying, composting, carbonizing and/or gasifying sludge. A part of compost and carbonized char are utilized as soil conditioner or lagging of molten steel in steel industries while the most of dry sludge, compost, char and gasification gas are recycled to biofuel in coal firing or renewal power generation. In those processes, drying of sludge becomes an important key technology to improve total energy efficiency because it consumes the largest energy (Stasta, et al., 2006, Zhang et al., 2009).

Several fundamental studies relating to sludge drying have been done on moisture sorption equilibrium (Vaxelaire, 2001, Ruiz and Wisniewski, 2008), drying induced shrinkage and crack formation (Chen et al., 2004, Tao et al., 2005, Hsu et al., 2010), skin layer formation (Tao et al., 2006, Vaxelaire et al., 2002, Font et al., 2011), structure change (Tao et al., 2006). There are also researches on drying kinetics of sludge in different dryers like fluidized dryer (Shin et al., 2000), convective drying (Leonard et al., 2005) and solar and heat pump dryer (Slim et al., 2008). In recent year, R&D on drying of sludge by composting has been carried out (Zhao et al., 2010). This drying process utilizes exothermic heat generated during aerobic degradation of organic compounds for evaporation of high moisture as alternative energy to heating from the outside. It looks a promising self-energy process to remove water from sludge, but the drying rate is influenced by aerobic fermentation rate and air stream pattern in sludge layer. Reduction of drying and/or composting time and deodorization are important issues to treat promptly a large amount of sludge as well as to introduce it in urban area.

The present authors propose an innovative technology of sewage treatment system. The feature is summarized as: 1) drying and composting rate of sludge is enhanced by mixing with char, 2) The dried sludge is carbonized to produce char. 3) Yield char is used for deodorization and water disposal of BOD and COD reduction as well as mixing with sludge in composting/drying process. These processes can establish self-energy closed sewage disposal system to reduce energy consumption.

In the present work, drying and composting behaviors of sludge was studied to examine the effect of char addition. A fundamental drying experiment was carried out by hot air heating to confirm the enhancement effect of sludge drying due to char addition. The drying and composting rate of sludge and char mixture was determined by examination using a lab-scale of disposer (1 kg/day) and a pilot scale of down draft column whose volume was 50 m^3.

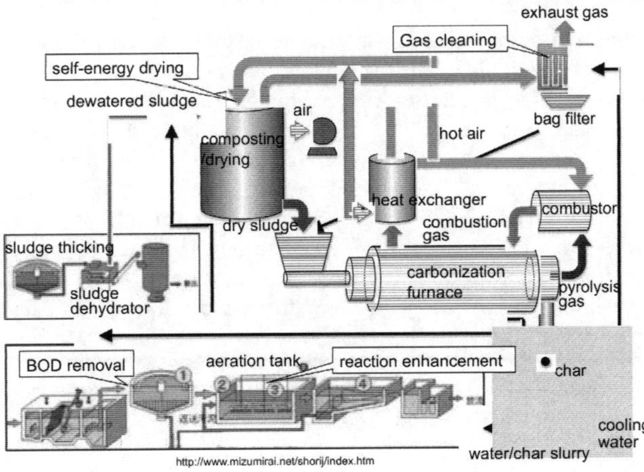

Fig. 1. Innovative self-closed sewage disposal and sludge carbonization system proposed

Enhanced Compost-Drying Process

Fig. 1 shows the outline of the enhanced sewage disposal and sludge carbonization system proposed in this study. Dewatered sludge separated from sewage is mixed with char powder and the mixture is subjected to composting in a down draft column. Then exothermic heat generated during composting induces simultaneously drying of sludge by air supplied to the column. The char particle may contribute to enhancement of composting and/or drying rate due to the effect of bacteria activation and water diffusion in an internal structure. Sequentially the dried sludge compost is decomposed to produce char in a kiln type of carbonization furnace by indirect heating using combustion of pyrolysis gas. The char is not only used in composting/drying processes of sludge, but also can be applied to agent for reduction of BOD and COD, enhancement of settlement of solid suspension in sewage treatment plant and gas cleaning by pre-coating on bag filter to remove ammonia and sulfur oxides causing offensive odor. Those techniques make it possible to establish self-energy closed treatment processes of sludge containing higher moisture content than 80% without heating from the outside and to reduce the total energy consumption in sewage disposal system.

Fundamental Drying Characteristic

Experiments

Fundamental drying behavior of sludge heated by hot air was examined preliminarily. A schematic drawing of a lab-scale of hot air drying examination setup is shown in Fig. 2. The setup consists of compressor, air reservoir, air heater and dryer. Stable hot air flow was maintained without fluctuation by using compressed air stored in reservoir from compressor. The flow rate of air was fixed at 0.35 m^3/min. The duct for hot air drying test was made of Styrofoam™ board with 30 mm thick, and the size was 100x100 mm in inner cross section and 1200 mm in length. Wire mesh was placed to distribute hot air at the inlet of the duct. The sample holder was a shallow cylinder made of plastics whose diameter and depth were 30 mm and 5 mm. The side and bottom of the holder were insulated thermally by Styrofoam, and only the top surface was subjected by hot air heating and evaporation of moisture. Dewatered sewage sludge used as the sample was provided from a sewage disposal center. Char produced by carbonization of sludge was mixed at different contents with sludge to examine the effect of char on drying behavior. The holder packed with sludge or sludge/char mixture was jacked up at the test section 1000 mm far from the inlet of the duct, and then drying experiment was started. The weight of the sample and holder was measured by balance and the drying rate was determined by the time behavior of the weight.

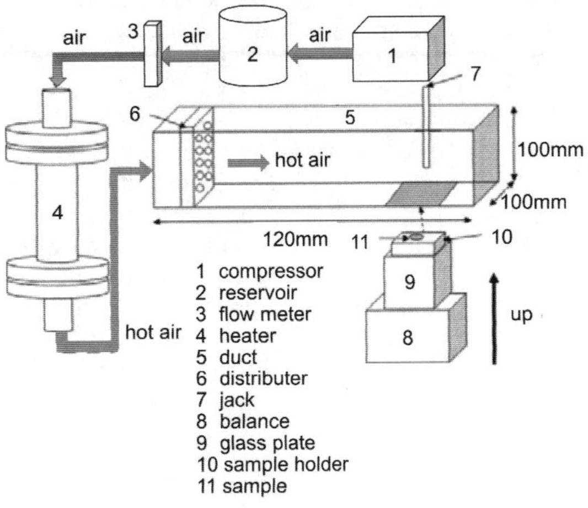

Fig. 2. A lab-scale of hot air drying setup

Drying characteristics

Time behavior of weight measured by balance was shown for different char contents in Fig. 3. The char content is defined by the ratio of char to initial wet sludge. The air temperature in this figure is 321 K. The curves look like that drying with higher char content completes in shorter time although the initial moisture content of each sample is different. In order to compare the drying behavior more in detail, drying characteristic curves were plotted on drying rate and moisture content. The moisture content is given by dry sludge basis here. The drying characteristic pattern showed no constant rate period in any char content because sludge forms immediately hard shell or skin on the surface as pointed out by the past references (Tao et al., 2006, Font et al., 2011) and it inhibits internal moisture diffusion. The curves reveal that higher char content results in greater drying rate through a whole drying period except lower moisture content of the sample with char content 0.05. This drying enhancement effect due to mixing char into sludge may advance moisture diffusion in the internal layer or shell layer formed by drying. The fast drying rate contributes to shortening of drying time. A similar drying characteristic was observed for drying experiment at another air temperature.

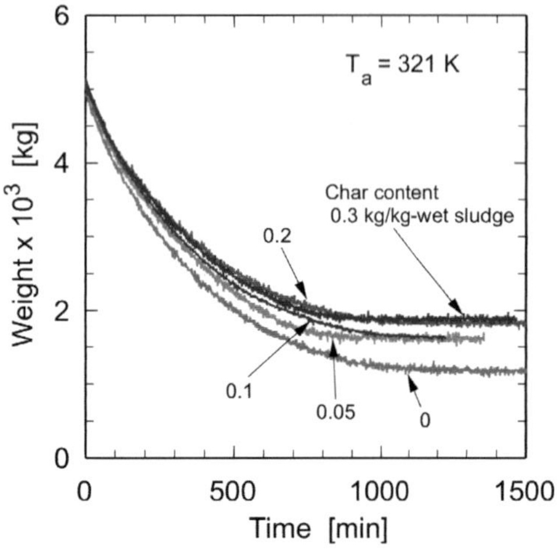

Fig. 3. Time behavior of sample with different char contents during hot air drying at 321 K

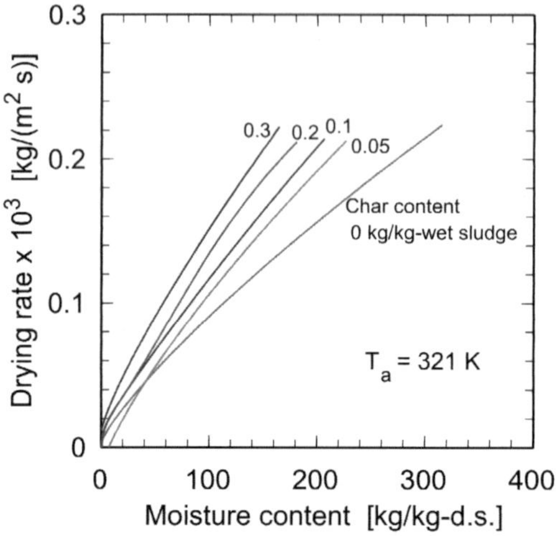

Fig. 4. Drying characteristic curves of samples for different char contents

Composting Characteristic

A lab-scale composting examination of sludge was carried out. The composting equipment used in this work is the one sold in market and the outline and picture are shown in Fig. 5. Bacteria seed for composting was preliminarily put into the vessel of the equipment. Composting rate depends on the kind of the bacteria seed, but the seed provided for composting of municipal wastes was used in this work as the most suitable bacteria was not known well and the main objective of this work is to study the possibility of drying only by exothermic heat in the composting step. Sludge or sludge/char mixture 1 kg was fed into the vessel in every day and the equivalent amount of compost to dry solid weight in the wet sample fed was taken out simultaneously. The moisture content of compost taken out was measured. These handling was repeated every day. The sample was stirred continuously by rotation blade during the examination.

Time behavior of moisture content during composting is compared between sludge and mixture of sludge and char in Fig. 6. The char content of the mixture was 0.11 kg-char/kg-wet sludge. The moisture content fell down gradually since beginning of experiment and achieved at steady state in 5 days for both the sample. It should be noted that the moisture content in the figure is given by the wet basis. The moisture content reduced to less than 30 % without any heating from outside. In this examination, little influence of char mixing was observed because the temperature and/or the composting rate in the vessel were too low to recognize the difference due to heat loss to the outside under cold atmospheric temperature in the winter season.

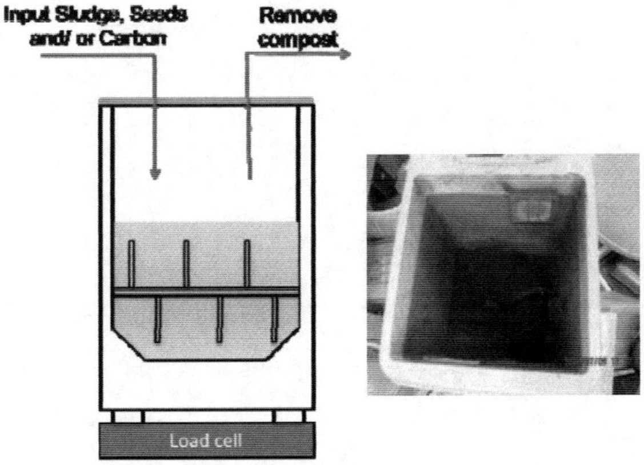

Fig. 5. Experimental setup of composting of sludge

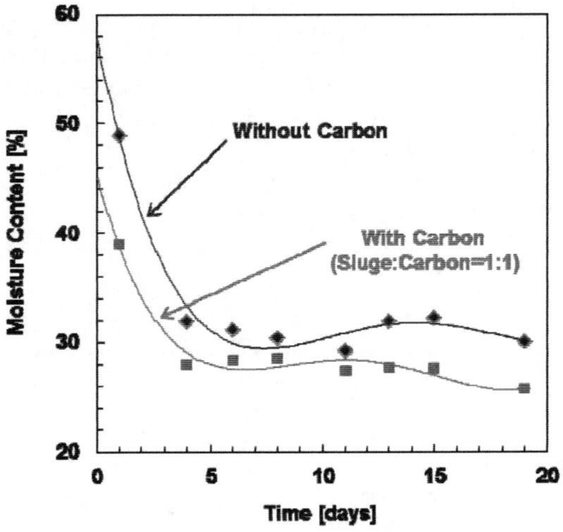

Fig. 6. Moisture content of sludge against time during composting examination

Conclusion

The innovative sewage disposal system was proposed to reduce total energy consumption by introducing self-energy closed processes for composting/drying and carbonization of sludge and char mixture. This study found the fundamental knowledge on an enhancement effect of char addition into sludge on drying and/or composting rate from hot air drying examination. In addition composting/drying could be performed with exothermic heat solely by aerobic decomposition and the rate was determined from the lab-scale and the pilot scale of examinations.

Acknowledgment

The present authors acknowledge New Energy and Industrial Technology Development Organization (NEDO) for financial supports to this study by "Research and Development Program for Innovative Energy Efficiency Technology (No. 10004805-0)".

Reference

1. Chen, J.B., X.F. Peng, T. Tao and D.J. Lee (2004), Thermal drying of wastewater sludge with crack formation, Water Sci. Technol., Vol. 50, pp. 177-182.
2. Hsu, J.P., T. Tao, A. Su, A.S. Mujumdar and D.J. Lee (2010), Model for sludge cake drying accounting for developing cracks, Drying Technology, Vol. 28, pp. 922–926.
3. Leonard, A., S. Blacher, P. Marchot, J.P. Pirard, M. Crine (2004), Measurements of shrinkage and cracks associated to convective drying of soft materials by X-ray microtomography, Drying Technology, Vol. 22, pp. 1695-1708.
4. Ruiz, T. and C. Wisniewski (2008), Correlation between dewatering and hydro-textural character-istics of sewage sludge during drying, Sep. Purif. Technol., Vol. 61, pp. 204-210.
5. Shin, Y.S., H.C. Kim and H.S. Chun (2000), Drying of water treatment process sludge in a fluidized dryer, Kor. J. Chem. Eng., Vol. 17, pp. 22-26
6. Slim, R., A. Zoughaib and D. Clodic (2008), Modeling of a solar and heat pump sludge drying system, Int. J. Refrig., Vol. 31, pp. 1156-1168.
7. Stasta, P., J. Boran, L. Bebar, P. Stehlik and J. Oral (2006), Thermal processing of sewage sludge, Appl. Therm. Eng., Vol. 26, pp. 1420-1426.
8. Tao, T., X.F. Peng and D.J. Lee (2005), Structure of crack in thermally dried sludge cake, Drying Technology, Vol. 23, pp. 1555-1568.
9. Tao, T., X.F. Peng and D.J. Lee (2006), Skin layer on thermally dried sludge cake, Drying Technology, Vol. 24, pp. 1047–1052.
10. Vaxelaire, J. (2001), Moisture sorption character-istics of waste activated sludge, J. Chem. Technol. Biotechnol., Vol. 76, pp. 377-382.
11. Vaxelaire, J. and J.R. Puiggali (2002), Analysis of the drying of residual sludge: from the experiment to the simulation of a belt dryer, Drying Technology, Vol. 20, pp. 989–1008.

12. Zhang, D.Q., P.J. He, L.Z. Yu and L. M. Shao (2009), Effect of Inoculation time on the bio-drying performance of combined hydrolytic-aerobic process, Bioresour. Technol., Vol.100, pp. 1087-1093.
13. Zhao, L., W.M. Gu, P.J. He and L.M. Shao (2010), Effect of air-flow rate and turning frequency on bio-drying of dewatered sludge, Waste Research, Vol. 44, pp. 6144-6152.

In -Situ Precipitation of $CaCO_3$ within Pulp Fibers Enhances Optical and Printing Properties of Paper

Pradeep Kumar, Yuvraj Singh Negi and Surendra Pal Singh

Department of Paper Technology, Indian Institute of Technology Roorkee, Saharanpur, India

Writing and printing papers are usually loaded with white pigment particles to improve many paper properties such as light scattering, smoothness, pore-size distribution, and ink-paper interactions. In the conventional method of loading, the filler particles are added directly into the papermaking stock before forming the paper sheet. In an alternative approach, the filler particles are placed within the lumen or the cell wall of the pulp fibers either by mechanical diffusion or by *in-situ* precipitation. Many researchers have found that the fiber-loading methods may offer several advantages over the conventional filler loading such as higher sheet strength, cleaner white water system, lower requirements of retention aids and improved formation of the sheet.

We have studied the effect of fiber-loading with $CaCO_3$ on paper sheets made from bagasse pulp. $CaCO_3$ particles were precipitated *in-situ* within the fiber cell wall and/or lumen by bubbling CO_2 gas through slurry of bagasse pulp fibers in a dilute aqueous suspension/solution of calcium hydroxide. Standard laboratory sheets were prepared from the filler-loaded pulp fibers. Strength and optical properties of the sheets were determined using standard test methods. The sheets were printed in the laboratory using an IGT printability tester. It was observed that the fiber loading by *in-situ* precipitation of $CaCO_3$ very significantly improved the light scattering coefficient of the bagasse pulp. As a consequence, the print through of the paper was decreased, which is particularly desirable in low grammage papers.

INTRODUCTION

Mineral fillers have been used in papermaking to reduce the cost and to improve brightness, opacity as well as the printability of the paper. In conventional method of filler loading, the filler particles are added directly into the stock. In finished paper, these filler particles occupy the inter-fiber voids in the sheet. Any attempt to increase filler content leads to reduced strength properties of the paper, cloudy paper machine back water, and dusting of paper during printing. Many of the drawbacks of conventional filler loading can overcome if the filler particles are placed within the fiber lumen or cell wall instead of the inter-fiber voids in paper sheet.

Methods for incorporating fillers inside fiber lumen or/and fiber cell wall have been studied extensively. Green et al. (1982) and Middleton and Scallan (1985) studied loading of fiber lumens with titanium dioxide particles by mechanical diffusion, in which the slurry consisting of softwood fibers and titanium dioxide was stirred over a period of time. Many other investigators followed similar approach using precipitated calcium carbonate (PCC) in place of titanium dioxide (Middleton et al. 2003; Miller and Paliwal 1985; Petlicki and van de Ven 1994).

Cellulosic fibers consist of a porous wall surrounding a central cavity (lumen). An estimated volume of pores in fibers is ≈ 1.5 cm^3/g fibers (Li et al 1995) and volume of the lumen is ≈ 0.2-0.4 cm^3/g fiber (Satyanarayana and Wypych 2007). Average size of the pores in the cell wall of unbleached softwood kraft pulp fiber is about 0.08 μm and that of the lumen is about 8 μm (Park et al. 2006). The various dimensions suggest that by mechanical dispersion the filler loading will be mainly into the fiber lumen and limited by the lumen volume. Therefore, the approach of lumen loading by mechanical diffusion is limited to rounded pigment particles having small enough particle size to pass through the pit apertures of the fibers to reach the lumen (Allan *et al.* 1992a). Alternatively, the filler particle could be incorporated within the cell wall voids (as high as 1.5 mL/g of pulp) by *in-situ* precipitation of insoluble inorganic materials. This concept of fiber loading or cell wall loading has been studied further by many researchers (Allan *et al.* 1992a, 1992b, 1998; Silenius 1996; Yoshida *et al.* 1987; Klungness et al. 1996, 2000; Subramanian et al. 2005; Chauhan et al. 2007; Kumar et al. 2009, 2011).

Although substantial disagreement exists in the findings of these researchers, the general view is that the paper made of pulp in which the filler has been incorporated within fibers exhibits greater tensile, burst, and tear strength than corresponding conventionally filled paper (Allan 1995; Miller and Paliwal 1985; Fujiwara 1993). Several other advantages of *in-situ* precipitation (Allan 1995) include easier white water management due to improved retention, reduction in the dosages of expensive polymeric retention aids, easier drying of paper due to reduced collapse of fiber voids, decreased abrasion damage to the forming fabric, less dusty paper, and smaller two-sidedness in paper because the filler is inside the fiber.

Klungness et al. (1996, 1999, 2000) found that the brightness and whiteness of pulp loaded by *in-situ* precipitation of calcium carbonate was less than that of the conventionally loaded pulp. Subramanian et al. (2005) from the study of blend of pulp and PCC composite fillers observed that adding PCC composite filler to the furnish enhanced light scattering properties of paper along with the tensile strength and Scott bond strength when compared with the conventional addition of PCC.

In the present work we have studied the effect of *in-situ* precipitation of $CaCO_3$ in bleached bagasse chemical pulp fiber on the optical and printing properties of the resulting paper sheets. It was observed that at equal ash levels, the bagasse pulp with calcium carbonate precipitated on fibers *in-situ* showed a significantly greater improvement in light scattering coefficient than precipitated calcium carbonate filler directly added to the pulp. The *in-situ* precipitation of filler also showed the desirable increase in print density and decrease in print through of printed sheets.

EXPERIMENTAL WORK

The pulps used in this study were never-dried bleached bagasse and hardwood pulps obtained from an integrated pulp and paper mill. The pulps were beaten in a PFI mill to 32 °SR. Calcium carbonate was precipitated *in-situ* within the pulp fibers by first saturating the pulp with calcium hydroxide solution and then carbonizing with pressurized carbon dioxide gas. 20 g pulp fibers with a calculated amount of a reagent grade calcium hydroxide powder were mixed with water in a beaker and stirred at 1200 rpm for 25 minute. The mass of calcium hydroxide was calculated to obtain a 15% $CaCO_3$ loading assuming its 100% conversion to $CaCO_3$, the resulted pulp. The amount of water added to the pulp was determined to obtain a desired consistency of 2% to 8% for the carbonizing reaction. After thorough mixing, the pulp slurry was transferred to a 2.5-L reaction vessel. The vessel was connected to a CO_2 gas cylinder, allowing flow of the gas into the vessel and obtaining a predetermined pressure (2-5 bar). After the reaction for a predetermined time, (3-12 minute), at the ambient temperature (≈ 30 °C), the vessel was depressurized. The pulp slurry was taken out from the vessel and washed thoroughly with running tap water over a 250-mesh screen until a clear filtrate was obtained to remove any $CaCO_3$ precipitated on the external surface of the fiber. The amount of precipitated $CaCO_3$ was determined by combustion of the pulp sample at 900 °C following the TAPPI T-413 standard method.

Standard hand sheets of 60 g/m^2 were prepared from the *in-situ* precipitated bagasse and hardwood pulps in a laboratory sheet former with a square cross section, 165mm x 165mm, according to the standard method SCAN C: 26. The hand sheets were air dried in contact with glaze plates. The hand sheets were internally sized by AKD to give $Cobb_{60}$ sizing test values of about 22 g/m^2. For comparison, standard handsheets were also prepared with direct addition of a commercial filler grade precipitated calcium carbonate (PCC-60) pigment and a retention aid (Percol47) obtained from a local chemicals supplier.

The brightness, opacity, specific scattering coefficient, and specific absorption coefficient of handsheets were determined from the measurements on Elerpho spectrophotometer (L&W Model 071). The other properties such as ash content and grammage of the handsheets were determined using appropriate SCAN methods. The physical properties of the pulps used in the study are given in Table 1.

Table 1. Physical Properties of the Pulps Used

Property*	Bagasse	Hardwood
Ash, %	0.61	0.66
Tensile index, kNm/kg	58.9	65.6
Burst index, kPa m^2/g	3.73	4.73
Tear index, mNm^2/g	3.96	5.34
Folding endurance	1.39	1.47
Brightness, %	73.5	73.0
Scattering coefficient, m^2/kg	21.2	36.7
Absorption coefficient, m^2/kg	0.52	0.73

The morphological characteristics of the fibers were determined using the L&W Fiber Analyser. The average fiber properties for bagasse and hardwood fibers are given in Table 2.

Table 2. Fiber Morphology

S. No.	Property	Bagasse	Hardwood
1.	Pulp freeness, °SR	32	32
2.	Mean fiber length (length weighted), mm	0.769	0.552
3.	Mean width (length weighted), μm	22.2	16.0
4.	Fiber Coarseness, μg/m	122.3	82.5
5.	Mean shape, %	88.3	92.6
6.	Number of vessels per 100,000 fibers	5245	264
7.	Number of vessels per gram, g^{-1}	452,705	35,628
8	Vessel area / fiber volume, μm^{-1}	0.000935	0.000182
	Mean vessel area, μm^2	155936	162724

Printing of handsheets

Handsheets of bagasse and hardwood pulps both loaded conventionally and by in-situ-precipitation were printed in an IGT printability tester (Model AIC_25). Test strips of the required length of 315 mm were prepared by butt-joining two handsheets with a thin adhesive tape at the backside. The paper strips were printed at varying amount of ink present on the printing disk while maintaining the other printing conditions constant as shown in Table 3.

A sufficient amount of ink was initially applied to the IGT inking unit and distributed uniformly on different rolls of the inking unit. During a test print, the printing disc was weighed, using an analytical balance with a least count of 0.1 mg, before and after the inking, and again after the printing. From these weights, it was possible to determine the amount of ink present on the printing disc before the printing and the amount of ink transferred to the paper during the printing.

The printing disc was cleaned after printing each strip using a solvent (cyclohexane). No fresh ink was applied to the inking unit after the initial application. This ensured that the amount of ink applied to the printing disk gradually decreased with printing of each strip. For each type of paper sheets, a series of strips was printed with varying amounts of ink on the printing disc. The initial amount of ink applied to the inking unit was adjusted such that the printing disk received anoint of ink equivalent to about 30 g/m^2 of the disc surface in the first inking. This amount decreased gradually with continued printing of paper strips. The prints were allowed to air-dry overnight. For calculation of print density and print through, various reflectance values on the test strips were measured using L&W Elerpho-071/070 brightness tester.

Table 3. Printing conditions in IGT printability tester

Atmospheric conditions during printing	(Temp=27±1°c; RH=65% ±5%)
Printing ink	Standard black offset ink supplied by the IGT. The ink was of the non-drying, oil-based type containing carbon black as pigment.
Printing speed	1 m/s
printing force	300 N
Printing disk	IGT Rubber covered disk, 50 mm
Backing material	None

RESULTS AND DISCUSSION

Effect of Filler on Optical Properties of Handsheets

Fig. 1 shows the effect of increasing amount of filler on the specific scattering coefficient, brightness, and opacity of the bagasse and hardwood handsheets. At zero filler level, the scattering coefficient of bagasse handsheets was much lower than that of hardwood handsheets. Direct filler loading resulted in marginal increase in the scattering coefficient for both bagasse and hardwood handsheets, while the fiber loading by *in-situ* precipitation resulted in a much greater increase, particularly for the bagasse handsheets. Both the bagasse and the hardwood pulps had nearly the same values of the scattering coefficient for ash content greater than about 12%.

The brightness and opacity of the handsheets made of bagasse fibers loaded by *in-situ* precipitation showed a very significant improvement because of the large increase in its scattering coefficient. To eliminate the effect of variation in the grammage of the handsheets, the values of opacity used in Fig. 1 were calculated for a grammage of 60 g/m^2 using the values of s and k, and the simplified Kubelka-Munk equations (Biermann 1996).

Offset printing behaviour of bagasse paper sheets loaded by in-situ-precipitation

Print density and print through were determined for the paper strips printed using different values of ink on the printing disk. Print density is the optical contrast between the printed and unprinted areas on the paper surface. Print through is an undesirable phenomenon in printing, in which the print is visible from the reverse side of the printed paper. The print density and the print through are usually determined from the measurement of reflectance values of printed and unprinted areas from the front and reverse sides of a print (SCAN P-36):

$$PD = \log (R_{\infty P}/R_P)$$

$$PT = 100 \times \log \left(\frac{R_{\infty B}}{R_{pB}} \right)$$

Where,

PD = print density

PT = print through

$R_{\infty P}$, $R_{\infty B}$ = reflectance of unprinted paper with a backing of opaque pad of the same paper on the printed side and on the reverse side respectively

R_P, R_{PB} = reflectance of the printed paper backed with an opaque pad of the unprinted paper on the printed side and on the reverse side respectively

The reflectance measurements were made on an L&W Elrepho-074 instrument corresponding to ISO Y-C/2°. An average of five measurements was recorded for each printed strip.

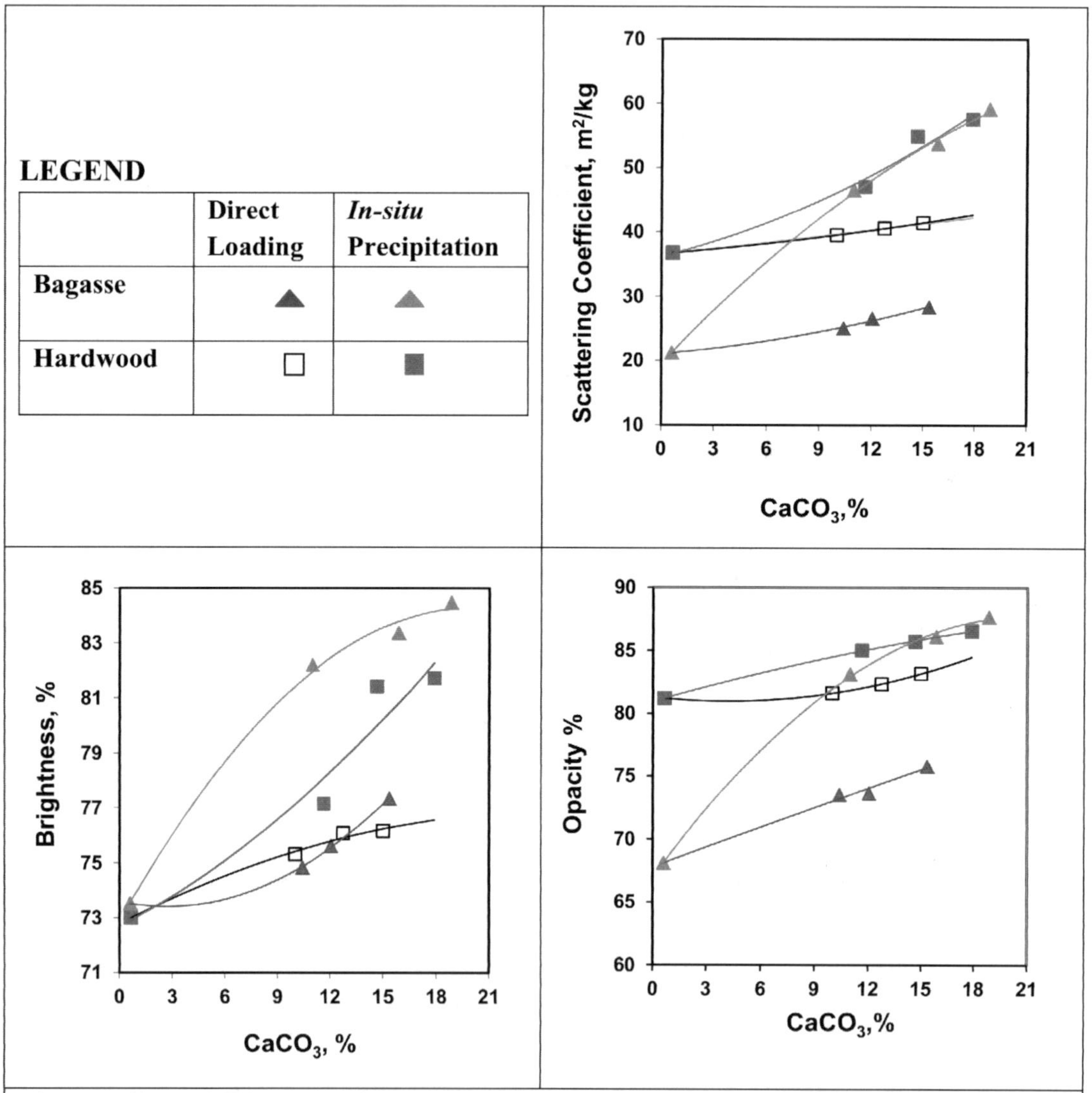

Fig. 1. Effect of filler on optical properties of the handsheets at different $CaCO_3$ levels. Opacity values were calculated for a constant grammage of 60 g/m^2 using simplified Kubelka-Munk equation.

Fig. 2 shows the variation of print density with varying values of amount of ink on the printing disk (x) and amount of ink on the paper (y) for both the bagasse and the hardwood handsheets. Here, x and y are expressed as g of ink per m^2 of disk or print area. An observation of the figures revealed the following points:

1. For any given ink transferred to the paper, the print density was greater for bagasse sheets than for the hardwood sheets.
2. For hardwood pulp, the print density at given amount of ink transferred to the paper was nearly the same for unfilled sheets as well as for loaded sheets, either by direct loading or by *in-situ* precipitation.
3. For bagasse pulp, the print density at given amount of ink transfer was different for unloaded and loaded sheets. The print density was lower of the loaded sheets than for the unloaded sheets. Further, the print density was more for the sheets loaded by *in-situ* precipitation than the sheets loaded directly.

Fig. 3 shows the variation of print through values for bagasse and hardwood sheets as function of x and y. For hardwood pulp, the print through decreased with loading of filler, but with little difference in the manner of filler loading, whether direct loading or loading by *in-situ* precipitation. On the other hand, for bagasse sheets, the effect of filler addition in reducing the print through was very significant, and more so for the loading by *in-situ* precipitation.

By increasing the amount of ink on the disk or the paper, the desirable improvement in print density is obtained along with the undesirable increase in the print density. Fig. 4 shows the print through as function of print density for bagasse and hardwood sheets. It was observed that for a given print density, the bagasse sheets loaded by *in-situ* precipitation had nearly the same values of print through as the hardwood sheets at equal filler level.

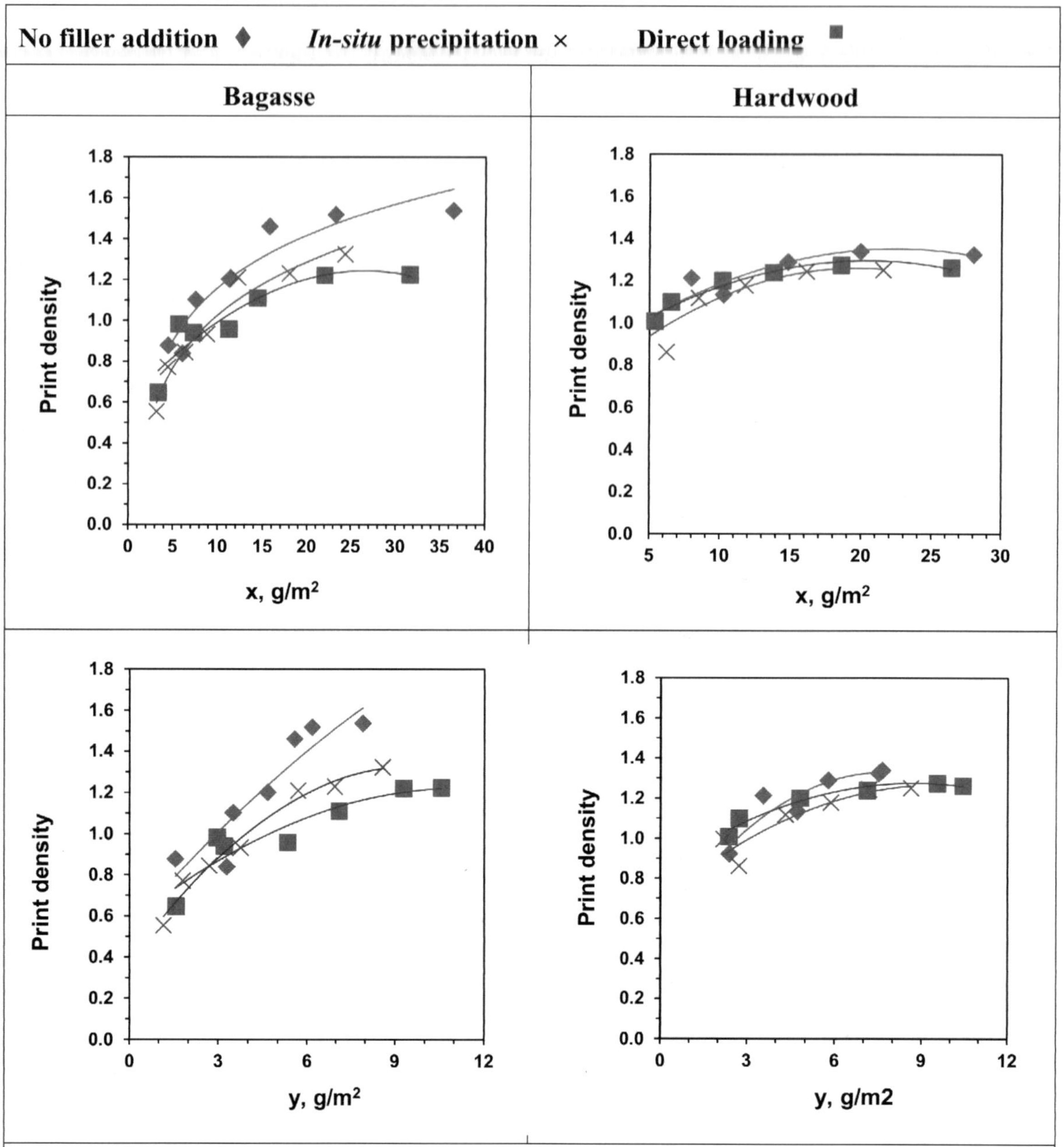

Fig. 2. Print density as a function of amount of ink on the printing disc and the amount of ink transferred to the paper.

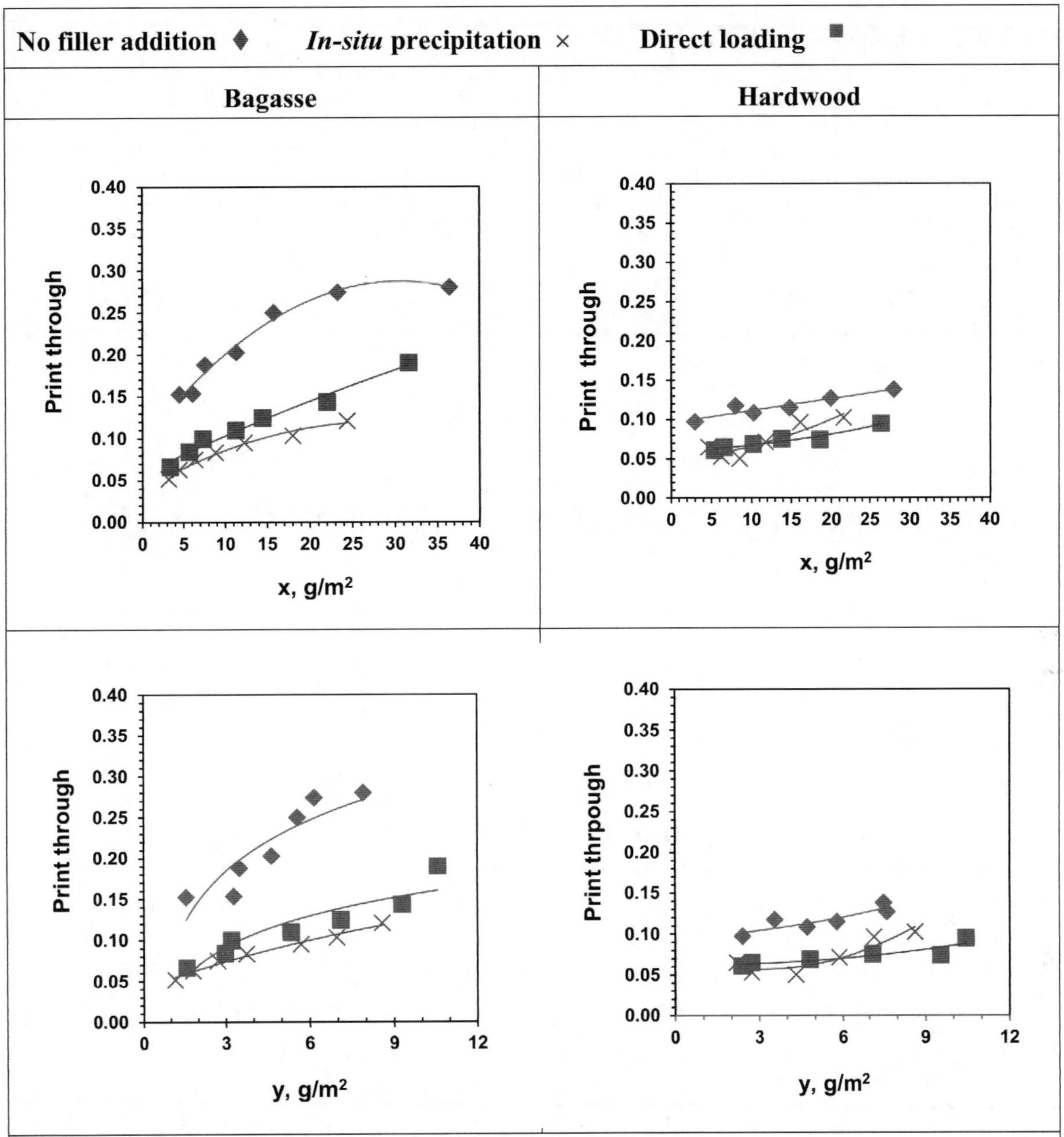

Fig. 3. Print through as a function of amount of ink on the printing disc and the amount of ink transferred to the paper.

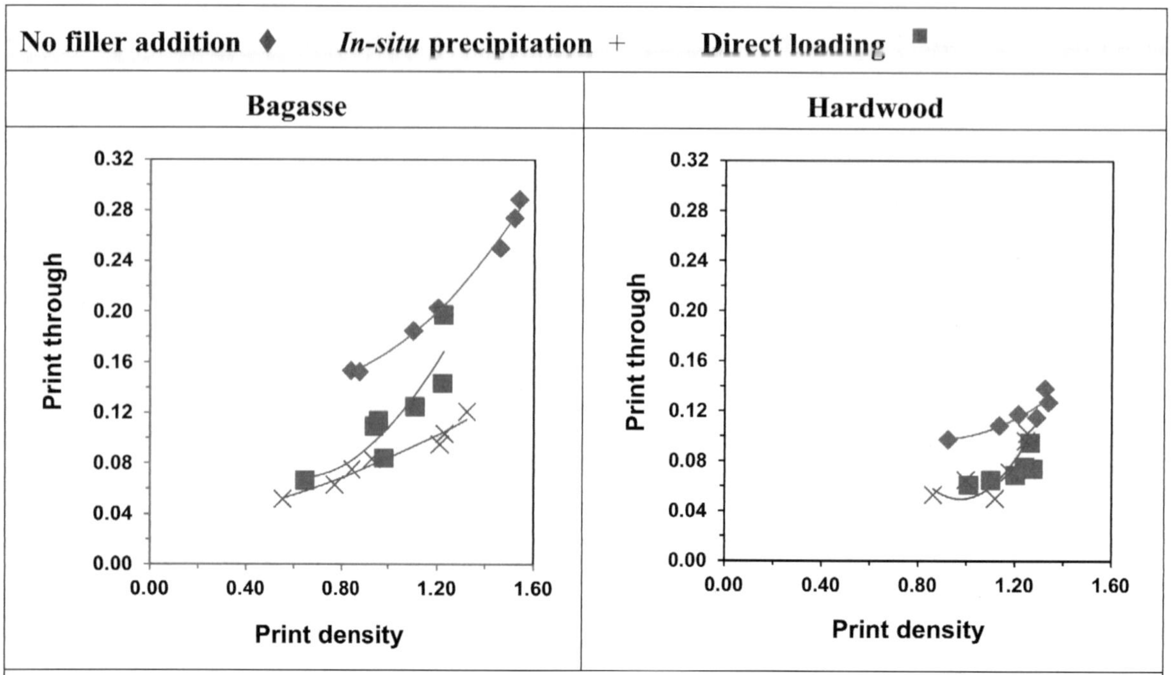

Fig. 4. Print through as a function of print density.

CONCLUSIONS

1. *In-situ* precipitation of calcium carbonate on bagasse fibers resulted in a very significant increase in specific scattering coefficient and consequently large improvements in opacity and brightness of the handsheets made of such pulp.
2. At the same level of filler loading, the scattering coefficient of *in-situ* precipitated pulp was much greater than that of the directly loaded pulp.
3. The improvement in optical properties of pulp by *in-situ* precipitation of calcium carbonate on pulp fibers was significantly higher for bagasse pulp than for the hardwood pulp.
4. *In-situ* precipitation of calcium carbonate caused a drop in strength properties of bagasse pulp. At equal ash level, the strength loss was more for loading by *in-situ* precipitation than by the direct filler loading. The strength loss of *in-situ* precipitated bagasse pulp could be recovered to a large extent by refining the in-situ precipitated bagasse pulp once again.
5. During offset printing, the ink transferred to paper sheet increased on loading of the paper with filler. However, the ink transfer was less for papers loaded by in-situ precipitation than for direct loaded papers. This difference in ink transfer behaviour could be due to difference in the positioning of the filler particles. In direct loaded sheets, the filler particles, present mainly in the inter-fiber voids, increase surface smoothness and fineness of the porous structure to facilitate easy ink transfer. On the other hand, in sheets loaded by *in-situ* precipitation, the filler particles present inside the fiber have little influence on these properties.

6. For *in-situ* precipitation of filler in bagasse fiber showed favourable effects on both the print density and the print through. For a given amount of ink on the printing disk or on the paper, the print density was more and the print through was less for *in-situ* precipitation of filler when compared with the direct filler loading

REFERENCES CITED

1. Allan, G. (1995). "Paper for the 21st Century," *Paper Age* (1), 10-15.
2. Allan, G., Carrol, J. P, Jimeney, G., and Negri, A. R. (1998). "Enhancement of the optical properties of a bagasse newsprint furish by fiber-wall-filler," *Cellulose Chem. Technol.*, 32, 339-347.
3. Allan, G., Carroll, J. P., Negri, A. R., Raghuraman, M., Ritzenthaler, P., and Yahiaoui, A. (1992a). "The microporosity of pulp: The precipitation of inorganic fillers within the micropores of the cell wall," *Tappi J.* 75(1), 175-178.
4. Allan, G., Negri, A. R., and Ritzenthder, P. (1992b). "The microporosity of pulp: The properties of paper made from pulp fiber internally filled with calcium carbonate," *Tappi J.* 75(3), 239-244.
5. Biermann, C. J. (1996). *Handbook of Pulping and Papermaking*, 2nd Ed., Academic Press.
6. Chauhan, V. S, Singh, S. P., and Bajpai, P. K. (2007) "Fiber loading of hardwood by *in-situ* precipitation of aluminosilicate," *BioResources*, 2(4), 560-571.
7. Fujiwara, K. and Morikawa, M. (1992). "Preparation of Magnetically Lumen-Loaded Pulp by In Situ Chemical Reaction", *Proc. Pan-Pacific Pulp Pap. Technol. Conf. (Tokyo), Sept. 8-10*,Part B, 131-134.
8. Green, H. V., Fox, T. J., and Scallan, A. M. (1982). "Lumen-loaded paper pulp," *Pulp & Paper Canada* 83(7), T203-T207.
9. Klungness, J. H., Ahmed, A., Ross-Sutherland, N., and AbuBakr, S. (2000). "Lightweight, high-opacity paper by fiber loading: filler comparison," *Nordic Pulp and Paper Research Journal,* 15(2), 345-350.
10. Klungness, J. H., Caulfield, D. F., Sachs, I. B., Sykes, M. S. and Tan, F. (1993). "Method for fiber loading a chemical compound," *U.S. Patent* 5, 223, 090.
11. Klungness, J. H., Pianta, F., Stroika, M. L., Sykes, M., Tan, F., and AbuBakr, S. (1999). "Lightweight high opacity paper: process cost and energy use reduction," AIChE *Symposium Series*, 95(322), 99-102.
12. Klungness, J. H., Syhes M.S., Tan,F.,Abubakr,S.,and EisenNassen,J.D.,(1996) "Effect of fiber loading on paper properties," *Tappi J.*79(3)297-301.
13. Kumar P, Gautam, S. K., Kumar, V., and Singh, S. P. (2009). "Enhancement of optical properties of bagasse pulp by *in-situ* filler precipitation," *BioResources* 4(4), 1635-1646.
14. Kumar, P., Negi, Y. S., and Singh, S. P. (2011). "Offset printing behavior of bagasse and hardwood paper sheets loaded by *in-situ*," *BioResources* 6(1), 207-218.
15. Middleton, S. R. and Scallan, A. M. (1985) "Lumen-loaded paper pulp: Mechanism of filler-to-fiber bonding," *Colloids and Surfaces* 16, 309-322.

16. Middleton, S. R., Desmeules, J., and Scallan, A. M. (2003). "Lumen loading with calcium carbonate fillers," *JPPS*, 29(7), 241-246.
17. Miller, M., and Paliwal, D. (1985). "The effects of lumen loading on strength and optical properties of paper," *J. Pulp Paper Sci.* 11(3), 84-88.
18. Petlicki, J., and van de Ven, T. G. M. (1994). "Kinetics of lumen loading," *J. Pulp Paper Sci.* 20(12), 375-382.
19. Silenius, P. (1996). "Preparation of filler containing papermaking materials by precipitating calcium carbonate *in-situ* in the presence of cellulosic materials," *Licentiate Thesis:* Lappeenranta University of Technology, Department of Chemical Technology, Lappeenranta, Finland.
20. Subramanian, R., Maloney, T., and Paulapuro, H. (2005). "Calcium carbonate composite fillers," *Tappi J.* 4(7), 23-27.
21. van de Ven, T. G. M. (1989). "Physicochemical and hydrodynamic aspects on fines and filler retention," In: Baker, C. F. and Punton, V. W. (eds.), *Fundamentals of Papermaking*, Transactions of the ninth fundamental research symposium held at Cambridge, Mechanical Engineering Publications Limited, London, Vol. 1, 471-494.
22. Yoshida, T., Kaneko, K., and Kazumori, K. (1987). "Manufacture of paper in neutral condition," *Japanese Patent* 62-162098.

(575a) Overview of Nanocellulose Process Operations

Mohini Sain, Center for Biocomposites and Biomaterials Processing, Faculty of Forestry, University of Toronto, Toronto, ON, Canada

An overview of nanocellulose will be presented with special emphasis to raw materials source, manufacturing process, performance and energy correlations. Presenttion will also briefly address the commercialization opportunities and challenges associated with nanocellulose and nanocellulose products.

Towards Co-Production of Composite Materials and Bioenergy: Effects of Enzymatic Saccharification On the Crystallinity, Reactivity, and Porosity of Woody Biomass

Han-Seung Yang[1], William T. Y. Tze[1], Jonathan Schilling[1] and Shona Duncan[1]

[1]Department of Bioproducts and Biosystems Engineering
University of Minnesota
Saint Paul, MN 55108-6130 USA
Tel: 612-625-5200 Fax: 612-625-6286
e-mail: hsyang@umn.edu

The objective of this study is to investigate the effects of different pretreatments and enzymatic saccharification levels on the crystallinity, reactivity, and porosity of solid residuals of woody biomass. Aspen wood flour was pre-treated with three different pretreatment agents, i.e., dilute acid, sodium hydroxide and peracetic acid, followed by different levels of enzymatic saccharification. The functional groups of solid residuals were analyzed using hydroxyl numbers to examine evidence of internal structural and chemical changes of the samples after each treatment in terms of hydroxyl group availability. The porosity and crystallinity of the solids were also evaluated using, respectively, nitrogen adsorption method (BET) and X-ray diffraction. The cell wall surface area, pore size, and pore volume will be determined from the BET method to discern the accessibility of the pre-treated substrates to enzymatic hydrolysis and also of the subsequent saccharification residues to value-adding conversion. X-ray diffractograms of samples subjected to different pretreatment processes and enzymatic hydrolysis levels will be evaluated. Crystallinity index, percent crystallinity, and cellulose crystallite size will be estimated for discussion of the implications to composite co-products to be produced from the saccharification residues.

(575c) Carbon Coated Nanocellulose Conductive Films for Flexible Electronic Applications

Yin Li[1], Makoto Schreiber[2], Stefano Gregori[1], Singaravelu Vivekanandhan[3], Amar K. Mohanty[4] and **Manju Misra**[3], (1)School of Engineering, University of Guelph, Guelph, ON, Canada, (2)Bioproducts Discovery and Development Centre, Department of Plant Agriculture, Crop Science Building, University of Guelph, Guelph, ON, Canada, (3)School of Engineering and the Department of Plant Agriculture, University of Guelph, Guelph, ON, Canada, (4)Department of Plant Agriculture & School of Engineering, University of Guelph, Guelph, ON, Canada

In recent years, there has been an increasing interest in the preparation of thin conductive films due to their lightweight and flexibility for applications in a wide range of electronic devices, packages, and connections. Current environmental concerns require industries, particularly the electronics industry, to find alternative methods and materials for the fabrication of various essential components. In addition, for commercial viability, the alternative materials must be inexpensive and the methods of fabrication must be simple and cost effective.

In our work, nanocellulose, a renewable and green material has been chosen as the base material for the films due to its strength and availability. In order to fabricate conductive films, the cellulose nanofibres were mixed with nanosized carbon-black particles. Both components were dispersed in water and made into films via a film casting method. For best results, the carbon particles should not just be dispersed but coated onto the cellulose fibres. The highly conductive films and insulating pure cellulose films can be fabricated layer by layer to build variable electrical capacitors which can be used as electrostatic transducers and energy harvesters. The morphology, conductivity, and durability of the films is being optimized. The films are characterized using scanning electron microscopy (SEM), transmission electron microscopy (TEM), dynamic mechanical analysis (DMA), and thermogravimetric analysis (TGA) for morphological, mechanical, and thermal analysis. The film conductivity is characterized using a Keithley 2701 multimeter to perform four probe measurements.

This research is financially supported by the Ontario Ministry of Agriculture, Food, and Rural Affairs (OMAFRA)-University of Guelph- (Bioeconomy for Industrial Uses Program); OMAFRA-New Directions Research Program 2009; the Ontario Research Fund (ORF) Research Excellence (RE) Round-4 from the Ontario Ministry of Economic Development and Innovations (MEDI).

Drug Loading in Pulp Cellulose Fiber and its Release Behavior

V. K. Rastogi, JoggGi Ganesh Dattatreya, Sujay Chattopadhyay, A. K. Ray

Department of Paper Technology, IIT Roorkee, Saharanpur 247 001, India

Abstract

In this present investigation, we are proposing cellulosic fibers (natural origin), to be used as drug carrier. Among the drug molecules sodium bicarbonate (an antacid) and paracetamol (a neutral anti-inflammatory) were selected to load into the cellulose fibers. Drug loading was carried out in cellulosic fibers by standard technique. Drug loaded fibers were properly washed to remove any surface adhered drug molecule. Then these drug loaded fibers were dipped into release medium with a constant stirring at three different temperatures i.e. 30°C, 35°C, and 40°C. Drug released from the fibers into phosphate buffered saline at above mentioned temperatures were estimated at various time intervals by measuring the absorption maxima of the drug by using the UV- spectrophotometer. From these measurements and the use of a mathematical model, we were able to determine the diffusion coefficients for sodium bicarbonate and paracetamol at various temperatures and finally activation energy values (12.8 kJ/mol for $NaHCO_3$, and 19.8 kJ/mol for paracetamol) were determined .Therefore, this extensive investigation was undertaken to examine the hypothesis whether the cellulosic fibers (natural) can be used as a effective drug carrier. In order to fulfil the aim a suitable mathematical model was used based on the geometry of the drug carrier and appropriate initial and boundary conditions.

1. Introduction

There are numerous synthetic and natural biodegradable polymers that have been investigated as drug carriers during last decade whose controlled drug release have attracted much attention due to advantages over conventional forms of dosage which decreases the toxic side effects by reducing the drug release rate and prolonging the release time [1]. An ideal drug delivery carriers should be made biocompatible, biodegradable, nontoxic material which minimizes drug degradation and loss, prevent harmful side-effects, increase the drug bioavailability and delivery at a specific target in the body.

Cellulose fibers are polysaccharides made up of glucose units and is considered a popular sustainable material as drug carrier. These are abundant, cheap, and safe and prevents drug from physiological condition. They can be easily modified with desired surface characteristics [2]. Reasonably high chemical stability, non-toxic, hydrophilic coupled with biocompatibility and biodegradability make it a popular option.

Moreover, polysaccharides are renewable resources and have high economical advantages over synthetic polymers also several studies have revealed that cellulose exhibited no significant cytotoxic effect and acts as a promising delivery vehicle for poorly soluble pharmaceuticals [3].

The transport properties of a novel local drug delivery system using nordihydroguaiaretic acid (NDGA)-polymerized collagen fibers [4] loaded with anti-inflammatory agents dexamethasone and dexamethasone 21-phosphate were studied. They proposed Fickian model and its application during a short time of diffusion give an estimate of this diffusivity and assumed that diffusion was the controlling mechanism for drug transport within the fiber. The developed equation showing the concentration of a diffusing substance from a long solid cylinder (Figure 1) of infinite length with a uniform distribution and non-steady-state conditions was of the

form [4] :

$$\frac{\partial c}{\partial t} = D\left[\frac{\partial^2 c}{\partial r^2} + \frac{1}{r}\frac{\partial c}{\partial r}\right] \qquad (1)$$

Where, c is the concentration of the drug in the fiber, t is the time of immersion, r is the radial distance within the fiber, and D is the diffusion coefficient of the drug in the fiber and they further assumed that (1)initially the drug distribution inside the fiber is uniform ($c = c_i$ at $t = 0$ for $0 < r < R$ where R is the radius of the fiber and c_i is the initial concentration); and (2) at the surface of fiber drug concentration is throughout to be zero ($c = 0$ for $t > 0$ at $r = R$). The series solution of the above equation is [4]:-

$$C(r,t) = \frac{2Ci}{R}\sum_{n=1}^{\infty}\frac{J_0(\alpha_n r)}{\alpha_n J_1(\alpha_n r)}\exp(-\alpha_n{}^2 Dt) \qquad (2)$$

The amount of diffusing substance that has left the cylinder in time *t* per unit area, M_t is [4]

$$M_t = -\int_0^t D\left(\frac{\partial c}{\partial r}\right)_{r=R} dt \qquad (3)$$

If M_0 is the initial amount of drug that was present in the fiber at t =0 then [5]

$$\frac{M_t}{M_0} = \frac{4}{R}\left(\frac{Dt}{\Pi}\right)^{0.5} \qquad (4)$$

Using the above equation the diffusion coefficient D, can be determined from the slope of the plot of $\frac{M_t}{M_0}$ vs. $t^{1/2}$ (cumulative drug release versus the square root of time).

Based on the existing literature, it can be inferred that to better understanding of the drug release profile we need to understand the interaction of the drug molecules with the carrier (cellulose here) and solubility of drug in the release medium, solution viscosity, temperature etc. Our knowledge is limited even with simple molecules and its interaction with cellulose fibers. In this paper we are proposing two classes of drug molecule which can be taken up for loading and release investigation of (i) basic salt (used an antacid) and (ii) neutral organic molecules (used as analgesic) and suitable theoretical model may be developed, that can predict the release behaviour.

2. Experimental

2.1 Materials

Softwood fibers (Pinus Roxburghii, 4 mm length and 40 μm diameter) were collected from nearby and integrated pulp and paper mill located in Northern India. Sodium bicarbonate ($NaHCO_3$) and methanol (CH_3OH) were purchased from M/S Fisher laboratories India. Paracetamol ($C_8H_9NO_2$) was purchased from Poklar pharmaceuticals Saharanpur India. For the production of phosphate buffer saline (PBS) sodium chloride (NaCl), potassium dihydrogen orthophosphate (KH_2PO_4), di-sodium hydrogen orthophosphate (Na_2HPO_4) and potassium chloride (KCl) are required and these all were purchased from M/S Fisher laboratories India.

2.2 Sodium bicarbonate loading

A weighted amount (2 gm) of cellulosic fibers were placed into 100 ml stock solution of $NaHCO_3$ (5 mg/ml) and incubated for 3 hours. The fibers were filtered out from the solution using filter paper followed by subsequent washing with phosphate buffered saline (PBS; pH 7.4) to remove any surface adsorbed drug molecule. The washed fibers were subsequently put inside oven at 105°C for 1 hr to obtain dry drug loaded fibers. After drug loading the fibers were divided into 8 equal (nearly) parts of 0.25 gm each and 7 parts were immersed in 25 ml of PBS buffer under continuous stirring using magnetic stirrer at three different temperatures i.e. 30°C, 35°C, and at 40°C for a specified time period (2, 4, 8, 16, 30, 60 min). After each of these time periods, aliquots (5ml) of the PBS was removed and analyzed by UV- spectrophotometer. The initial amount of drug that was present in the fibers was estimated by taking out the 8th part of fibers in 50 ml of fresh PBS and gently heated at 60°C for 30 min to ensure maximum removal of all drug molecules present in the fibers.

2.3 Paracetamol loading

A weighted amount (1.5 gm) of cellulosic fibers were placed into 100 ml stock solution of paracetamol (20 µg/ml) and incubated for 3 hours. The stock was prepared in the methanol but not in water as the solubility of paracetamol in methanol is much greater than water. The fibers were filtered out from the solution using filter paper followed by subsequent washing with phosphate buffered saline (PBS; pH 7.4) to remove any surface adsorbed drug molecule. The washed fibers were subsequently put inside oven at 60°C for 1 hr to obtain dry drug loaded fibers. After drug loading the fibers were divided into 6 equal (nearly) parts of 0.25 gm each and 5 parts were immersed in 50 ml of PBS buffer under continuous stirring using magnetic stirrer at three different temperatures i.e. 30°C, 35°C, and at 40°C for a specified time period (5,15,30,45and 60 min). After each of these time periods, aliquots (5ml) of the PBS was removed and analyzed by UV- spectrophotometer. The initial amount of drug that was present in the fibers was estimated by taking out the 6th part of fibers in 50 ml of fresh PBS and gently heated at 60°C for 30 min to ensure maximum removal of all drug molecules present in the fibers.

2.4 In vitro drug release of Sodium bicarbonate and Paracetamol

Amount of $NaHCO_3$ and paracetamol from the collected aliquot was estimated through UV spectrophotometer at wavelength of λ=210nm (absorption maxima for sodium bicarbonate) and λ =249.5nm (absorption maxima of paracetamol) through a previously formed calibration curve in PBS at their respective same wavelength.

3. Results and discussion

3.1 Sodium bicarbonate Loaded Fibers

By dissolving different amounts of sodium bicarbonate in PBS (pH 7.4) a standard calibration curve for sodium bicarbonate in PBS was determined and a linear correlation were obtained between absorbance and concentration. The sodium bicarbonate loaded fibers were incubated in PBS at three different temperatures i.e. 30°C, 35°C, and 40°C. After the specified incubation periods, the aliquots were analyzed on UV- spectrophotometer for sodium bicarbonate content. A relationship between the sodium bicarbonate absorbance versus time was found. For each sample, the absorbance obtained was converted to the equivalent concentration using the standard calibration curve. A percentage drug release versus time plot was determined (Figure 2), from the resulting concentration versus time data.
Analysis of (Figure 2) shows, a burst release of drug occurs during first 10 min and there after the release is steady. Temperature change of the release medium does not influence the initial burst period, while the release after 10 min becomes reasonably slow. From the slope of the curve in (Figure 3), the diffusion coefficient of sodium bicarbonate at 30°C in the cellulosic fibers was found using eq 4 to be D = 0.919 x 10-9 m^2/sec and likewise for rest of the temperatures (Table 1).

3.2 Paracetamol Loaded Fibers

By dissolving different amounts of paracetamol in PBS (pH 7.4) a standard calibration curve for paracetamol in PBS was determined and a linear correlation was obtained between absorbance and concentration. The paracetamol loaded fibers were incubated in PBS at three different temperatures i.e. 30°C, 35°C, and 40°C. After the specified incubation periods, the aliquots were analyzed on UV- spectrophotometer for paracetamol content. A relationship between the paracetamol absorbance versus time was found. For each sample, the absorbance obtained was converted to the equivalent concentration using the standard calibration curve. A percentage drug release versus time plot was determined (Figure 4), from the resulting concentration versus time data. Analysis of (Figure 4) shows, a burst release of drug occurs during first 10 min and there after the release is steady. Temperature change of the release medium does not influence the initial burst period, while the release after 10 min becomes reasonably slow. From the slope of the curve in (Figure 5), the diffusion coefficient of sodium bicarbonate at 30°C in the cellulosic fibers was found using eq 4 to be D = 0.985 x 10^{-10} m^2/sec and likewise for rest of the temperatures (Table 1).

3.3 Discussion of drug release phenomena

Analysis of (Figure 2 and Figure 4) shows that the paracetamol release is slower than $NaHCO_3$, but the % release is not very much dependent on temperature of the release medium. Nearly identical pattern of release is found for both $NaHCO_3$ and paracetamol. After placing the dry drug loaded fibers into PBS buffer, fibers take up enough water and their pore structures swells which help in easy release of drug from the cellulose matrix. This leads to burst phenomena. At the subsequent stage, the release becomes slow and nearly attains equilibrium. This shows that drug-fiber interaction is not significant possibly due to extremely large pore size of fibers chosen here as carrier. This is yet to be established from further investigation of the fiber microstructures before and after drug loading and release. Nearly 60% of the loaded drug dissolves out during first 10 min for $NaHCO_3$, while it is nearly 30% for paracetamol under similar situation. This may be attributed to (i) the low solubility of paracetamol in PBS buffer and (ii) size of drug molecule probably causes this slight variation in burst effect. At steady state after 45min nearly 20-25% $NaHCO_3$ is retained in the fiber while, 35-40% drug is retained in case of paracetamol.

3.4 Estimation of Activation energy for sodium bicarbonate and paracetamol

Activation energy can directly tell about the diffusion barrier in the matrix for different kinds of drugs based on the several types of ionic interactions, size of the drug molecule, pore size of matrix etc.

It can be calculated by the following,

$$D = D_0 e^{-\frac{E}{RT}} \qquad (5)$$

$$\ln D = \ln D_0 - \frac{E}{RT} \qquad (6)$$

From the slope of the curve in (Figure 6 and Figure 7), the activation energy for sodium bicarbonate and paracetamol were found using eq 6 to be E = 12.8 kJ/mol for $NaHCO_3$, and E = 19.8 kJ/mol for paracetamol.

4. Conclusions

In this paper, we investigated that cellulosic fibers can act as drug carrier, but its efficacy needs to be improved by additional surface treatment which will reduce the initial burst release. Although we have tested a salt of strong base-weak acid and a neutral paracetamol molecule, but these fiber can also be loaded with other classes of therapeutic drugs (i.e. acidic, basic and neutral). Obviously a thorough understanding of the stability of the drug loaded fiber needs to be estimated before we come to final decision. Possibly fiber pore size and the molecular interactions of drugs (chosen here) with fibers are weak which causes rapid initial burst. Therefore, these cellulosic fibers are suitable where there an initial burst release of drugs is desired. Diffusion of paracetamol is slightly slower than that of sodium bicarbonate as obtained from the (Figure 2 and Figure 4) as well as activation energy values (12.8 kJ/mol for $NaHCO_3$ and 19.8 kJ/mol for paracetamol). This is attributed to a larger size of paracetamol molecules and low solubility in PBS buffer compare to $NaHCO_3$. Thus, a proper surface treatment of the drug loaded fiber will certainly change this release behaviour.

Acknowledgements

The authors thankfully acknowledge to Staff members, Department of Paper Technology and Mr. Ashutosh Lamba, PhD Scholar for their support and guidance in order to complete the experimental work.

References

1. C. Peng, Q. Zhao, and C. Gao, Sustained delivery of doxorubicin by porous $CaCO_3$ and chitosan/alginate multilayer-coated $CaCO_3$ micro particles. Colloid and Surface A: *Physicochemical and Engineering Aspects*, 353, (2009)132–139.
2. H. Baumann, C. Liu, and V. Faust, Regioselectively modified cellulose and chitosan derivatives for mono- and multilayer surface coatings of hemocompatible biomaterials. Cellulose, 10, (2003) 65–74.
3. X. Wang and R. Sun, self assembled lignocelluloses micelles: A new generation of value added functional nanostructures *BioResources* 6(3), (2011) 2288-2290.
4. Y. Moussy, E. Guegan, T. Davis, and T. J. Koob Transport, Characteristics of a Novel Local Drug Delivery System Using Nordihydroguaiaretic Acid (NDGA)-Polymerized Collagen Fibers *Biotechnol.* Prog. 23, (2007) 990-994.
5. J. Siepmann, and N.A. Peppas, Modeling of drug release from delivery systems based on hydroxypropyl methylcellulose (HPMC), *Advanced Drug Delivery Reviews,* 48, (2001) 139–157.

Table 1: *Calculated diffusivities for sodium bicarbonate and paracetamol.*

Temp(°C)	**Diffusivity of Sodium**	**Diffusivity of Paracetamol(m^2/sec)**
30°C	0.919×10^{-9}	0.985×10^{-10}
35°C	1.000×10^{-9}	1.150×10^{-10}
40°C	1.084×10^{-9}	1.280×10^{-10}

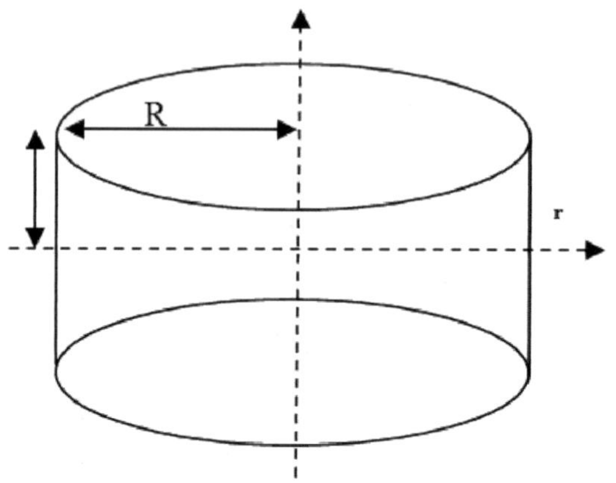

Fig 1: *Cylindrical geometry of fiber.*

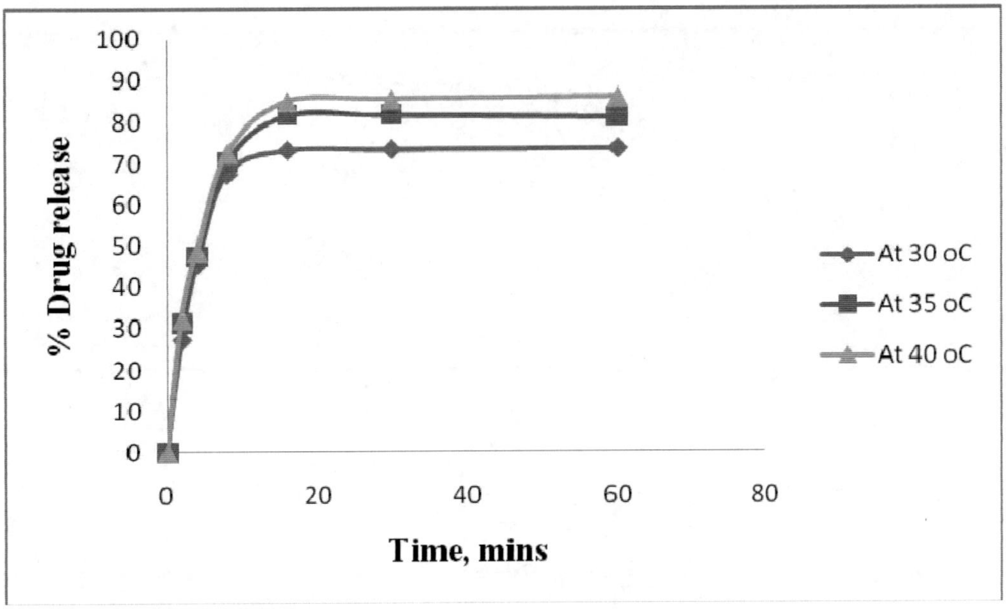

Fig 2: Release profile of sodium bicarbonate in PBS buffer at 30°C, 35°C and 40°C.

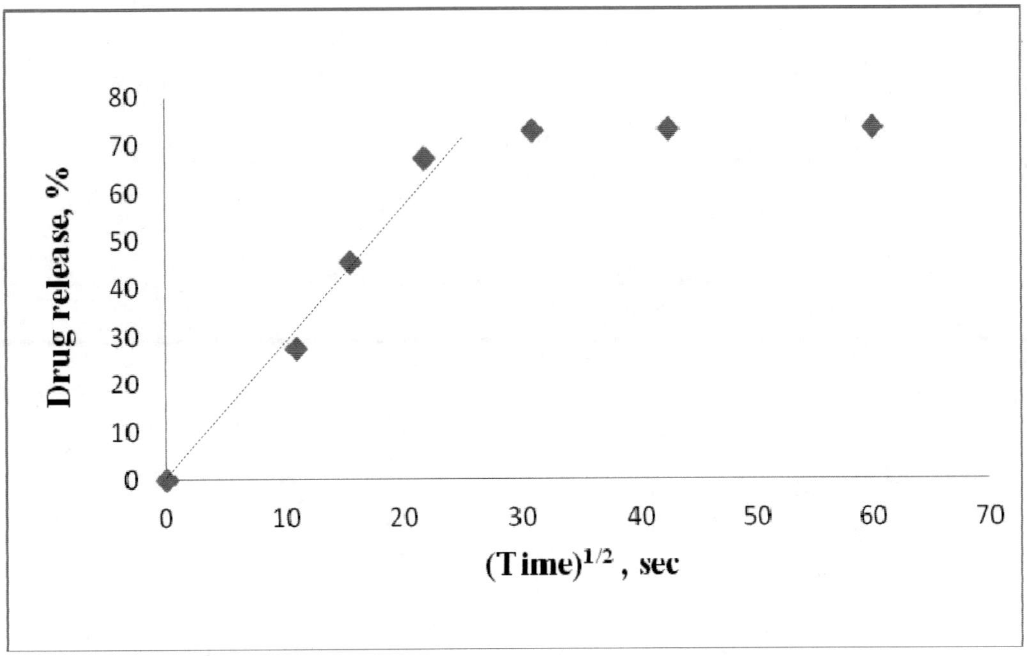

Fig 3: Percent release of sodium bicarbonate at 30°C into PBS versus the square root of time. Dashed lines indicate the best fit of eq 4 to the initial data points.

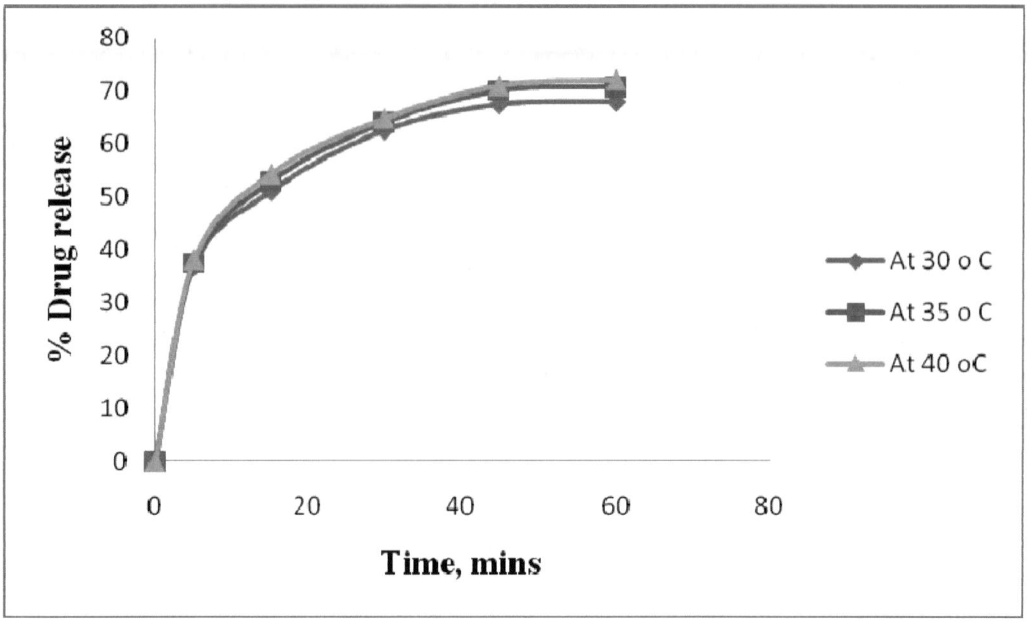

Fig 4: *Release profile of paracetamol in PBS buffer at 30°C, 35°C and 40°C.*

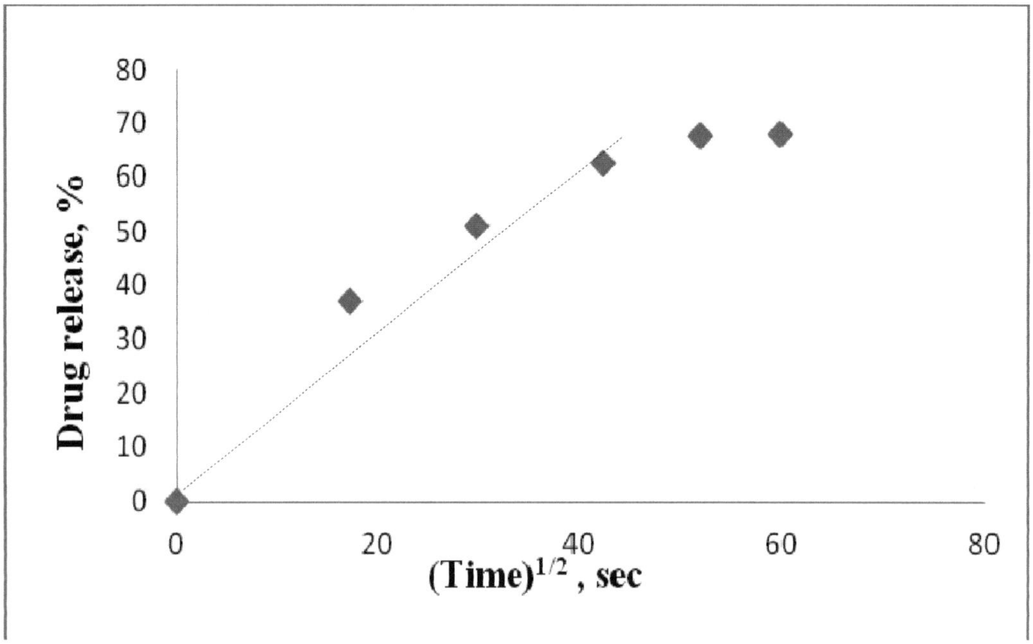

Fig 5: *Percent release of paracetamol at 30°C into PBS versus the square root of time. Dashed lines indicate the best fit of eq 4 to the initial data points.*

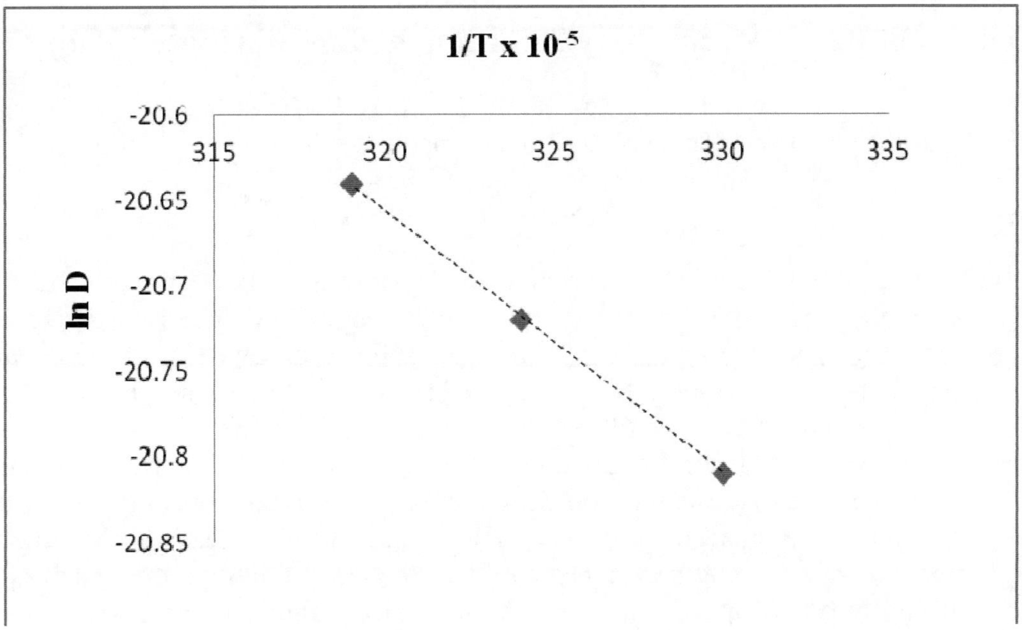

Fig 6: *Plot of 1/T vs. ln D at 30°C, 35 C and 40°C for sodium bicarbonate.*

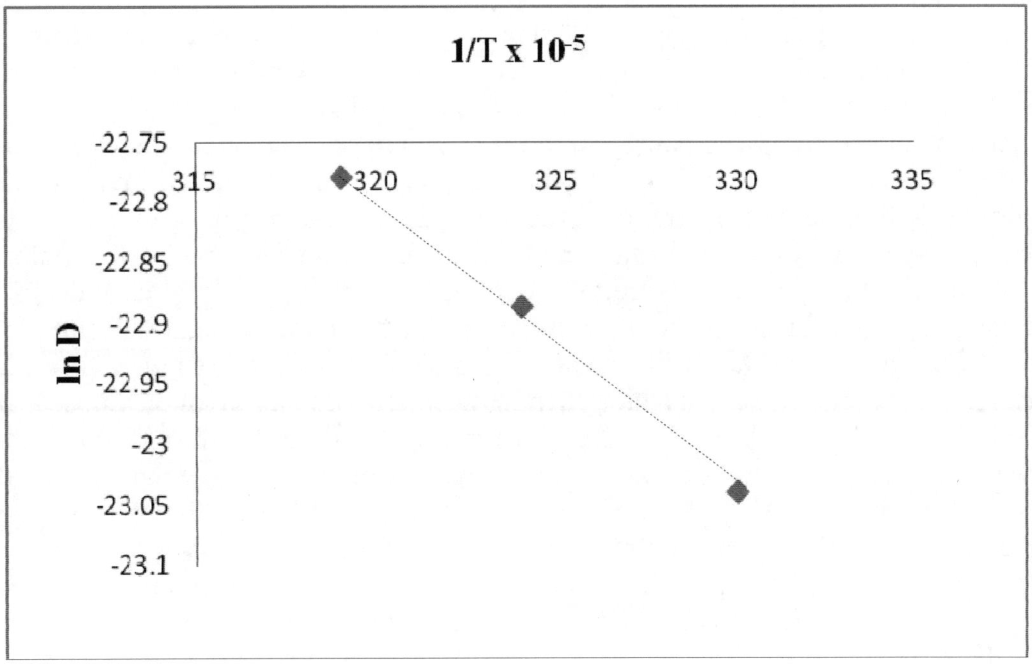

Fig 7: *Plot of 1/T vs. ln D at 30°C, 35 C and 40°C for paracetamol.*

Improvement of Paper Strength by Addition of Nanofibrillated Pulp Fibers

Arvind kumar, A. K. Singh and S. P. Singh
Department of Paper Technology, Indian Institute of Technology Roorkee,
Saharanpur Campus, Saharanpur (India) 247001

Abstract:

Nanofibrillated pulp fibers can be prepared by several methods such as chemical method (Acid or Alkali hydrolysis), high shear refining, ultrasonication and TEMPO mediated oxidation methods. In the present study high shear refining has been followed for the preparation of nanofibers from the bleached bagasse pulp. Prepared nanofibers were used in the papermaking process as an additive in different quantity varying from 5 to 20 percent. Hand sheet prepared by above method has been analyzed for different strength properties like tensile strength, tear strength, burst strength and optical properties like brightness and opacity. Hand sheets of bleached bagasse pulp with small quantities of nanofibrillated pulp fibers showed improvement in tensile strength and burst strength of paper with little difference in tear strength. The effect of high shear refining on the isolated nanofibers was studied using chemical analysis and field emission scanning electron microscopy.

Keywords: bagasse; high shear refining; nanofibers; hand sheets

Introduction

Cellulose is one of the most abundant biopolymers on earth. It is widely used as sustainable raw material for papermaking because of its economical availability, biodegradability, and biocompatibility. Chemically, cellulose is a polydispersed linear polymer of poly-β (1, 4) -D-glucopyranose with a syndiotactic configuration [1]. Cellulose is the most important constituent of the plant cells, and forms a structural framework around which all other components such as hemicellulose, lignin and pectin are deposited during the plant cell growth [2]. Cellulose microfibrils are synthesized by a plasma membrane and are a self-assembly of cellulose chains that are organized into a perfect sterioregular configuration through a regular network of inter and intra molecular hydrogen bonds. The fibers exhibit a unique structural hierarchy derived from their biological origin. The fibrils isolated from natural fibers normally have a wide range of diameters with most of them smaller than 100 nm [3]. Isolation of cellulose nanofibrils (CNFs), smaller than 100 nm has attracted a lot of interest in the last few years due to its unique characteristics such as high surface area, high tensile strength, high young's modulus, low coefficient of thermal expansion [3]. CNFs have shown great potential in several applications, including biomedical [4], bionanocomposites [5], oxygen gas barrier layers [6] and optically transparent functional materials [7]. CNFs have been extracted from a number of commercially exploited plant materials [8-15, 17]. CNFs can also be obtained from algae, tunicates, and bacterial cellulose [18, 19].

Many different processes have been used to obtain nanofibers from cellulosic materials. These methods include mechanical treatments, such as grinding [20], cryocrushing [21], and high pressure homogenizing [22], as well as chemical treatments such as acid hydrolysis [23], enzyme-assisted hydrolysis [24], TEMPO-mediated oxidation followed by a mild mechanical treatment [25]. Some other methods combine both mechanical and chemical treatments such as chemically ultrasonication process [26] and electrospinning methods [27]. All these methods lead to different types of nanofibrillar materials, depending on the cellulose raw material, pretreatment and disintegration process.

Hassan et al. prepared nanofibers using the chemical treatment and followed by mechanical method and made handheets of paper in the labotatory and observed that the paper made from nanocellulose fibers had wet strength about 12 times greater than the strength of normal papers [28]. Anna Lindgren and Sara Wennberg had been prepared the nanofibers using the TEMPO mediated oxidation followed by mild mechanical treatment and made handsheets of the paper in the laboratory and observed that as the level of nanofibers increase, the strength properties of the paper sheets increased [29] .

The aim of the present work was to isolate cellulose nanofibers from bleached bagasse pulp by mechanical refining and to study the effect of these nanofibers on the properties of the resulting paper. The chemical composition and morphology of the isolated nanofibers were characterized by means of chemical analysis and field emission scanning electron microscope (FE-SEM). The nanofibers produced from bleached bagasse kraft pulp were blended in different proportions with the normally refined pulp and were made into standard laboratory handsheets. The properties of the handsheets were evaluated using standard methods.

Experimental

Commercial grade bleached bagasse kraft pulp was used for the isolation of nanofibrils. The pulp sample received from a bagasse pulp mill was air dried in the laboratory and stored at room temperature. The pulp was analyzed for cellulose and hemicelluloses content using TAPPI method T 203 cm-99. The pulp had 86% cellulose and 14% hemicelluloses (Average of three measurements). The pulp fibers were beaten under high shearing conditions in a PFI mill. The PFI mill was charged 30 g of pulp fiber at 10% consistency for beating. One batch of pulp fibers was beaten for 3500 revolutions of the mill and another batch of the pulp was beaten for 100,000 revolutions. The freeness values of the beaten pulps, measured using the TAPPI standard T 227 om-99, were found to be 35 °SR for pulp beaten to 3500 revolutions and 90 °SR for pulp beaten to 100,000 revolutions. In a PFI mill, a cylindrical rotor with narrow axial bars on its curved surface rotates in a smooth housing. During the operation of the mill, the rotar is pushed to one side of the housing. The pulp fibers remain in contact with the inner surface of the mill housing due to the centrifugal force and are repeatedly subjected to high shear stresses as they are carried into the narrow gap formed between the housing and the rotor bars .

The pulps beaten to different freeness were examined using a field emission scanning electron microscope (FE-SEM, Make: FEI Quanta 200 F microscope) to observe the effect of refining on the fiber dimensions. A dilute suspension of pulp fibers and fibrils was spread on a glass slide in a thin layer and allowed to dry. These fibers and fibrils were coated with gold (Au) by an ion sputter coater. The coated sample was observed under the SEM operating at 12.5 kV.

The 35 °SR has beaten pulp was blended with 90 °SR has beaten pulp in different proportions; 5%, 10%, 15%, and 20% of 90 °SR pulp by weight. Standard handsheets of 60 g/m^2 were prepared from these blends on a laboratory sheet making machine (British Sheet Former) using the TAPPI standard method T 205 sp-02. The handsheets were wet pressed at 345 kPa (50 psig) and air dried in contact with gloss plates.The hand sheets were evaluated for tensile strength using method T 494 om-01, tear strength using method T 414 om-04 and burst strength using method T 403 om-02.

Results and Discussion

Figure 1 shows the FE-SEM micrographs of the pulp fibers beaten to 35 °SR and 90 °SR at 500X and 4000X levels of magnification.

	35 °SR (3500 PFI revolutions)	90 ° SR (100,000 PFI revolutions)
Magnification 500X	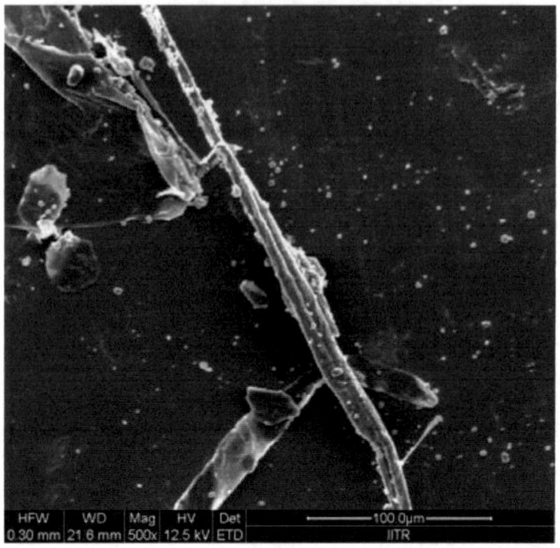	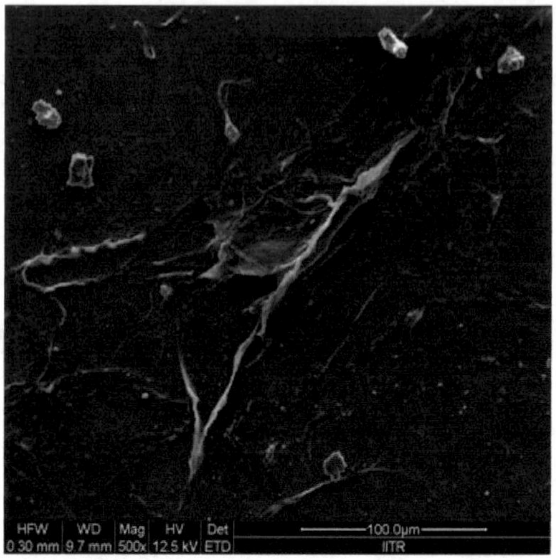
Magnification 4000X	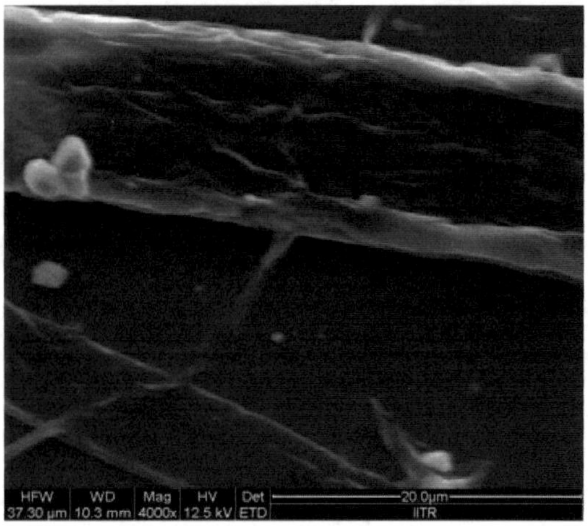	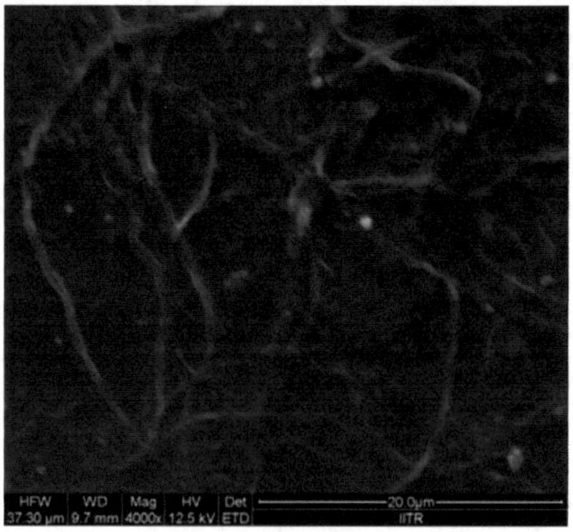

Figure 1: FE-SEM images of bagasse pulp fibers beaten to 35 °SR and 90 °SR.

The papermaking pulp fibers are normally beaten to a freeness of 35° SR. The refining has four primary effects on the fibers, namely, (1) internal fibrillation (breaking of intra-fiber bond), (2) external fibrillation (rupture of the primary layer of the cell wall and exposure of the secondary layer), (3) fiber shortening, and (4) production of fines. The micrographs show that the fibers beaten to 35 °SR largely maintain their fiber integrity, but a prolonged beating of pulp to 90 °SR leads to disintegration of the fibers into microfibrils / nanofibrils. Continued energy input during refining appears to be able to break the hydrogen bonds between the cellulose microfibrils and also the bond between microfibrils and hemicelluloses.

The micrographs at a magnification of 4000X show that the dimensions of the microfibrils/ nanofibrils produced on refining the pulp to 90 °SR were in the range of 200–600 nm width and a few tens of micrometer length.

Figure 2 shows the effect of blending of highly beaten pulp (90 °SR nanofibrillated) with normally beaten pulp (35 °SR) strength properties of the bleached bagasse pulp fibers.

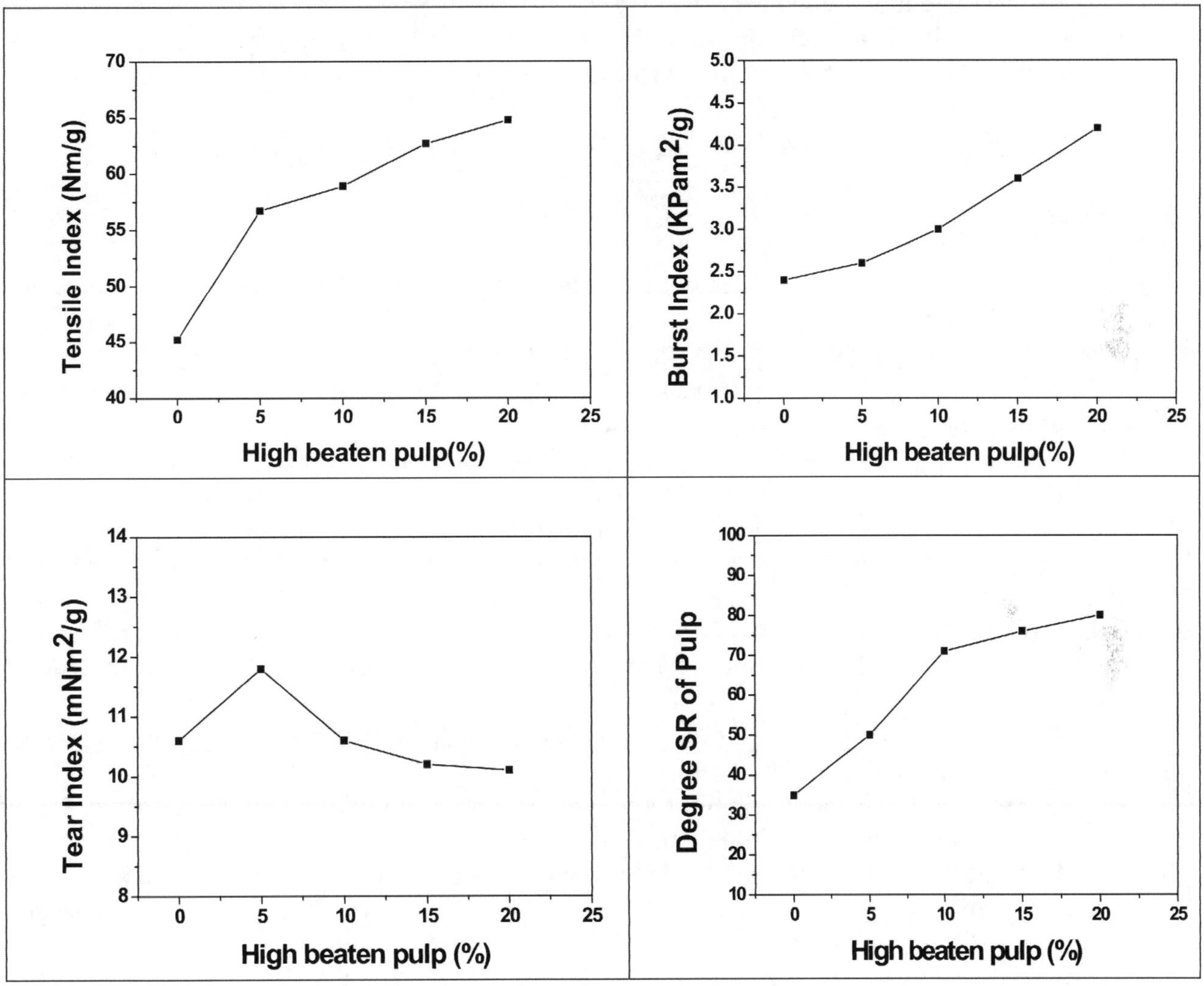

Figure 2: Effect of blending composiotion of highly beaten (90 °SR) pulp with normally beaten (35 °SR) pulp on strength properties.

The addition of nanofibrillated pulp to normally beaten pulp results in a significant improvement in the tensile index and burst index of the sheets, but has little influence on the tear index. These results imply that the small quantities of the highly beaten pulp (nanofibrils) enhance fiber-fiber bonding. However, the blending of the slow nanofibrillatred pulp causes a significant reduction in the freeness of the pulp mixture.

Conclusion

1. Prolonged refining of bleached bagasse pulp fibers resulted in the disintegration of the fiber into microfibrils/ nanofibrils in the dimension range of the range of 200–600 nm width and a few tens of micrometer length.
2. Blending of microfibrils /nanofibrils produced by prolonged refining with normally refined pulp results in a significant improvement in tensile and burst strength of the handsheets made from these pulps. The effect of blending is not very significant on tearing strength of the handsheets.
3. Blending of the slow nanofibrillatred pulp with normally refined pulp causes a significant reduction in the freeness of the pulp mixture, which would reduce the sheet dewatering papermaking.
4. Making pulps of such high °SR levels require very high refining energy.

References:

1. Nishino, T., Matsuda, I., & Hirao, K. (2004). All-cellulose composite. Macromolecules, 37(20), 7683–7687.
2. Clowes FAL, Juniper BE (1968) Plant cells. Blackwell Scientific Publications, Oxford
3. Habibi Y., Lucia LA., Rojas O.J. (2010) Cellulose nanocrystals: chemistry, self-assembly, and applications. Chem Rev 110(6):3479–3500.
4. Czaja WK., Young DJ., Kawecki M, Brown RM (2006) the future prospects of microbial cellulose in biomedical applications. Biomacromol 8(1):1–12.
5. Liu H., Liu D., Yao F., Wu Q. (2010) Fabrication and properties of transparent polymethylmethacrylate/cellulose nanocrystals composites. Bioresource Technol. 101(14):5685–5692.
6. Fukuzumi H., Saito T., Iwata T., Kumamoto Y., Isogai A. (2009) Transparent and high gas barrier films of cellulose nanofibers prepared by TEMPO-mediated oxidation. Biomacromol. 10(1):162–165.
7. Nogi M., Iwamoto S., Nakagaito AN., Yano H. (2009) Optically transparent nanofiber paper. Adv Mater 21(16):1595–1598.
8. Abe K., Iwamoto S., Yano H. (2007) Obtaining cellulose nanofibers with a uniform width of 15 nm from wood. Biomacromol 8(10):3276–3278.
9. Abe K., Yano H. (2010) Comparison of the characteristics of cellulose microfibril aggregates isolated from fiber and parenchyma cells of Moso bamboo (Phyllostachys pubescens). Cellulose 17(2):271–277.
10. Cherian BM., Lea˜o AL., de Souza SF, Thomas S., Pothan LA., Kottaisamy M. (2010) Isolation of nanocellulose from pineapple leaf fibers by steam explosion. Carbohydrate Polymer 81(3):720–725.
11. Chen Y, Liu C, Chang PR, Cao X, Anderson DP (2009) Bionanocomposites based on pea starch and cellulose nanowhiskers hydrolyzed from pea hull fiber: effect of hydrolysis time. Carbohydrate Polymer 76(4):607–615.
12. De Morais Teixeira E., Corre^a A., Manzoli A., de Lima Leite F., de Oliveira C., Mattoso L. (2010) Cellulose nanofibers from white and naturally colored cotton fibers. Cellulose 17(3):595–606.

13. Alemdar A, Sain M (2008) Isolation and characterization of nanofibers from agricultural residues-wheat straw and soy hulls. Bioresource Technol 99(6):1664–1671.
14. Wang B, Sain M, Oksman K (2007) Study of structural morphology of hemp fiber from the micro to the nanoscale.Appl Compos Mater 14(2):89–103.
15. Rosa MF, Medeiros ES, Malmonge JA, Gregorski KS, Wood DF, Mattoso LHC, Glenn G, Orts WJ, Imam SH (2010) Cellulose nanowhiskers from coconut husk fibers: effect of preparation conditions on their thermal and morphological behavior. Carbohydr Polym 81(1):83–92.
16. Zuluaga R, Putaux JL, Cruz J, Ve´lez J, Mondragon I, Gan˜a´n P (2009) Cellulose microfibrils from banana rachis: effect of alkaline treatments on structural and morphological features. Carbohydr Polym 76(1):51–59.
17. Dinand E, Chanzy H, Vignon RM (1999) Suspensions of cellulose microfibrils from sugar beet pulp. Food Hydrocolloid 13(3):275–283.
18. Iwamoto S, Kai W, Isogai A, Iwata T (2009) Elastic modulus of single cellulose microfibrils from tunicate measured by atomic force microscopy. Biomacromol 10(9):2571–2576.
19. Hirai A, Inui O, Horii F, Tsuji M (2008) Phase separation behavior in aqueous suspensions of bacterial cellulose nanocrystals prepared by sulfuric acid treatment. Langmuir 25(1):497–502.
20. Abe K, Yano H (2010) Comparison of the characteristics of cellulose microfibril aggregates isolated from fiber and parenchyma cells of Moso bamboo (Phyllostachys pubescens). Cellulose 17(2):271–277.
21. Chakraborty A, Sain M, Kortschot M (2005) Cellulose microfibrils: a novel method of preparation using high shear refining and cryocrushing. Holzforschung 59(1):102–107.
22. Nakagaito AN, Yano H (2004) The effect of morphological changes from pulp fiber towards nano-scale fibrillated cellulose on the mechanical properties of high-strength plant fiber based composites. Appl Phys A-Mater 78(4):547–552.
23. Liu, H., Liu, D., Yao, F.,&Wu, Q. (2010). Fabrication and properties of transparent polymethylmethacrylate/cellulose nanocrystals composites. Bioresource Technology,101 (14), 5685–5692.
24. Pa¨a¨kko¨ M, Ankerfors M, Kosonen H, Nyka¨nen A, Ahola S,O¨ sterberg M, Ruokolainen J, Laine J, Larsson PT, Ikkala O, Lindstro¨m T (2007) Enzymatic hydrolysis combined with mechanical shearing and high-pressure homogenization for nanoscale cellulose fibrils and strong gels. Biomacromol 8(6):1934–1941
25. Isogai A (2009) Individualization of nano-sized plant cellulose fibrils by direct surface carboxylation using TEMPO catalyst under neutral conditions. Biomacromol 10(7):1992–1996.
26. Tischer PCSF, Sierakowski MR, Westfahl H, Tischer CA (2010) Nanostructural reorganization of bacterial cellulose by ultrasonic treatment. Biomacromol 11(5):1217–1224
27. Frenot A, Henriksson MW, Walkenstro¨m P (2007) Electrospinning of cellulose-based nanofibers. J Appl Polym Sci 103(3):1473–1482.
28. Hassan M.L, Mathew A.P, Hassan E.A, Oksman K (2010) Effect of pretreatment of bagasse pulp on properties of isolated nanofibers and nanopaper sheets. J Wood and Fiber Science, 42(3):1-15.
29. Lindgren A, Wennberg S (2010) Preparation of Nanofibers from Pulp fibers (Thesis). Applied Surface Chemistry.

(603a) Cellulosic Ethanol Production From Ionic Liquid Pretreated Biomass with a Recombinant Yeast Expressing Cellulases

Nobuiro Ishida, Biotechnology Laboratory, Toyota Central R&D Labs., Inc., Nagakute, Aichi, Japan, Satoshi Katahira, Toyota Central R&D Labs., Inc., Nagakute, Aichi, Japan, Satoshi Saito, Toyota Motor Co., Toyota, Japan, Noriho Kamiya, Center for Future Chemistry, Kyushu University, Fukuoka, Japan and Chiaki Ogino, Graduate School of Engineering, Kobe University, Kobe, Japan

As the primary polysaccharide of plant cell wall, lignocellulose has received considerable attention as a main feedstock for bio-refinery process such as cellulosic ethanol production. The enzymatic hydrolysis of lignocellulose to soluble sugars is considered to be the most important process in the production of environmental friendly bioethanol. However, the spontaneous crystallization of cellulose due to the chemical uniformity of the glucosyl residues and the high degree of hydrogen bonding can result in densely packed micro fibrils that are inaccessible to cellulolytic enzymes. For the efficient utilization of cellulosic materials in biomass, there is a need to enhance the enzymatic saccharification of this biopolymer.

For deconstruction of biomass with low energy, ionic liquids were employed as pre-treatment media, and the effective degradation and assimilation procedures were investigated [1)]. Subsequently, we developed a transgenic *Saccharomyces cerevisiae* strain that was secretary produced cellulase, and examined the ethanol fermentation from ionic liquid pretreated biomass [2)]. The overall process has been named as 'Ionic liquid-based Consolidated Bio-Processing (i-CBP)'. In this study, we will be presented the effective cellulosic ethanol fermentation from ionic liquid pre-treated biomass using the transgenic *S. cerevisiae* producing the optimal ratio of saccharification enzymes.

1) Uju et al., *Bioresource Technol.*, Vol. 103, p 446 (2012).

2) K. Nakashima et al., *Green Chem.*, Vol. 13, p 2948-2953 (2011).

(603b) A Quantitative Determination of the Effects of Surface/Bulk Lignin and Xylan Contents On Enzymatic Hydrolysis of Lignocellulosic Biomass

Xiaohui Ju, Chemical Engineering and Bioengineering, Washington State University, Richland, WA, Mark Engelhard, Pacific Northwest National Laboratory , Richland, WA and Xiao Zhang, Voiland School of Chemical Engineering and Bioengineering, Bioproducts Science Engineering Laboratory, Washington State University, Richland, WA

Lignocellulosic biomass provides an abundant source of sugars for sustainable biofuels and biochemical production. However, the lack of highly efficient and cost effective cell-wall degrading enzymes is a major obstacle preventing the commercialization of a biologically based lignocellulosic biomass conversion process. The effectiveness of enzymatic hydrolysis is inextricably linked to the structural and chemical characteristics of the biomass material. Due to the diverse nature of lignocellulosic biomass and the heterogeneous structural and chemical characteristics of pretreated biomass substrates, it is difficult to investigate the effect of individual substrate characteristic on enzymatic hydrolysis. We have developed a new methodology based on chemical pulping principles to create a set of biomass "reference substrates" with controlled properties to reflect chemical and structural diversity of biomass materials. The individual effects of surface/bulk lignin and xylan contents on enzymatic hydrolysis of these substrates were determined.

X-ray photoelectron spectroscopy (XPS) has been demonstrated as an effective tool to investigate the surface lignin content and its effect on enzymatic hydrolysis. A Response Surface Methodology (RSM) was employed to attain an aggregated understanding of surface/bulk lignin and xylan contents influence on biomass recalcitrance.

(603c) Use of a Carbon Budget Model to Calculate Land Use Change Greenhouse Gas Emissions Associated with Forest-Based Biofuels Production in Michigan

Jiqing Fan[1], David R. Shonnard[1], Robert E. Froese[2] and Robert Handler[3], (1)Chemical Engineering, Michigan Technological University, Houghton, MI, (2)Michigan Technological University, Houghton, MI, (3)Sustainable Futures Institute, Michigan Technological University, Houghton, MI

Abstract:

Michigan is heavily forested with 19.4 million acres of timberland, and the large volume of timber resources are potential feedstocks for biofuel and bioenergy production. To truly understand the environmental impacts of biofuels production in Michigan, it is essential to evaluate the land use change (LUC) impacts associated with the biomass production practices in this area, namely sustainable harvesting from natural regeneration forest land, and short rotation forestry (SRF) plantation. These two practices are not likely to cause indirect land use change (iLUC) emissions because forest products are not used in the food market. However, the impacts of direct LUC (dLUC) should be examined. The Carbon Budget Model of Canadian Forest Sector (CBM-CFS3) is applied in this study to estimate the dLUC impact associated with the biomass production in Michigan. The CO_2 flux between the atmosphere and soils due to dLUC are quantified and incorporated into the LCA study of the biofuels. We consider three cases for this model; 1. Afforestation of MI forests, 2. Increased harvesting above a business as usual harvest level to satisfy a certain level of biofuel production, and 3. Establishment of SRF plantation on abandoned agricultural lands to satisfy a certain level of biofuel production. A business as usual (BAU) harvesting scenario is first established based on the current harvesting acreage and intensity, a case in which MI forests continue to accrue carbon as forests recover from historic harvest practices (Afforestation case). Because the carbon budget model simulates over a several decade time period, we will also include estimates of changes in fossil greenhouse gas emissions due to changes in extraction practices, for example by relying in inputs from high greenhouse gas intensity crude oils, such as tar sands, using literature values. SRF species, such as poplar, can be planted on abandoned or marginal agricultural land, but tillage at the land preparation stage will disturb the carbon balance and microbial activity in soil, while the growing biomass can sequester carbon from atmosphere into the root system as it grows. Finally, we include these landscape level carbon dynamics into life cycle assessments of biofuels that can be produced from forest feedstocks, for example cellulosic ethanol, to understand the potential impacts on product-based LCA results.

(603d) Gasification Characteristics of Methane Fermentation Residues and Aquatic Weeds

Takeshi Nakajima, Kiyoto Matsumoto, Masanori Ochi, Soshi Watanabe, Junichi Ida, Tatsuki Toda, Tatsushi Matsuyama and Hideo Yamamoto, Department of Environmental Engineering for Symbiosis, Soka University, Hachioji, Tokyo, Japan

In this study, pyrolysis and steam gasification of methane fermentation residues of food waste and aquatic weeds were performed by changing experimental conditions such as reaction temperature and water vapor concentration. In the experiment, the pyrolysis and steam gasification were carried out under two different heating rates (1K/s or instant temperature raise). In the slow heating experiment, time course change of the sample weight, evolved gases and generated tar and char were monitored and analyzed under inert or steam containing atmosphere for the temperature range of 473-973 K. As a result, it was found that under inner gas atmosphere, CO_2 evolution by primary pyrolysis and H_2, CO, CH_4 evolution by secondary pyrolysis were observed for all the samples used. On the other hand, under steam containing atmosphere, rapid H_2 evolution was observed after primary pyrolysis reaction. This may be due to the water-gas shift reaction by steam reforming.

In the rapid heating experiment, samples were inserted into the electric furnace which was pre-heated to desired temperature (873 ~ 1273K) and pyrolysis or steam gasification were performed under inert or steam containing atmosphere. At the exit of the furnace, all the effluent gases were collected to gas bag, and the total yields of the generated gases were examined using GC-MS and GC-TCD. As a result, high H_2 yield was observed from all samples under steam gasification condition. This may also be due to the water-gas shift reaction by steam reforming. In addition, with the increase of the reaction temperature, the yields of the total combustible gases also increased. Futhermore, among all the used samples, Elodea nuttallii (aquatic weed) showed the highest total combustible gases yields.

(603e) Microfiltration of Lignocellulosic Hydrolyzates Using Dead End Filtration to Separate Lignin and Other Inhibitors

Bhavin Bhayani, Empire State Paper Research Institute, Syracuse, NY

Microfiltration of lignocellulosic hydrolyzates using dead end filtration to separate lignin and other inhibitors

B. Bhayani, B.V. Ramarao

Department of paper & Bioprocess engineering, SUNY ESF

In the production of biofuels from lignocellulosics, the biomass is first extracted using dilute mineral acids or hot water. This hemicellulose rich oligomers are hydrolyzed under treatments at higher temperature and pressure to yield sugar monomers which are fermented into biofuels such as ethanol, biobutanol or bioplastics. The initial treatment with either dilute acids or hot water also allows some of the lignin to pass in particulate, colloidal or dissolved forms into the sugar solution. Lignin and other phenolic compounds derived from it are strongly inhibitory to microbial fermentation processes. They also adhere to polymeric membranes, carbohydrates and other surfaces, fouling separation membranes. Their removal from the liquor stream not only facilitates downstream separations but improves the fermentation profiles of the resulting sugar hydrolyzates. Therefore, methods to separate lignin from liquors and hydrolyzates form valuable additional operations in the biorefineries.

A new approach has been developed which can separate inhibitory components from the extracts based on the fact that reduction of temperature of the biomass hydrolyzates can result in the native agglomeration and separation of lignin containing solids by simple settling and/or filtration which thus can result in large increases in separation efficiencies for hydrolyzates. By filtering the extracts at lower temperature (10°C) it is possible to obtain higher flux rate than at higher temperatures (20°C). As shown in Figure 1 and Figure 2, the cumulative flow for extracts at 10°C is higher than the extracts at 20°C at two different pressures. The turbidity of permeate for any given combination of pressure and temperature was less than 10 NTU. It has been hypothesized that the natural flocculation tendency of extracts increases with decreasing temperature.

In addition to the temperature, the effect of addition of a flocculent (Polyethylene oxide, PEO) has been investigated. Substantial enhancement in filtrate flux upon addition of 20 ppm of PEO as seen in Figure 3 is observed.

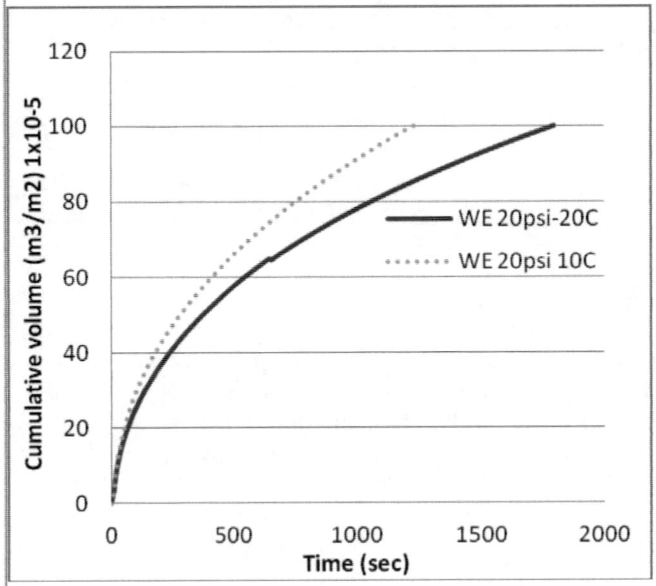

Figure SEQ Figure * ARABIC 1: Cumulative flow at different temperature at 20psi

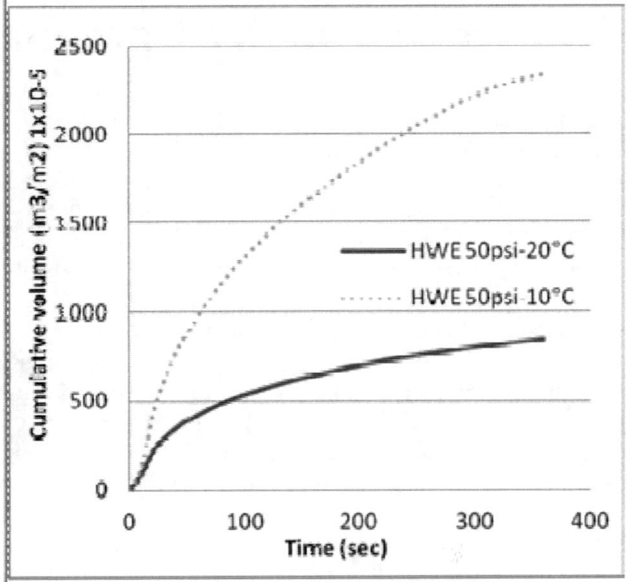

Figure SEQ Figure * ARABIC 2: Cumulative flow at different temperature at 50psi

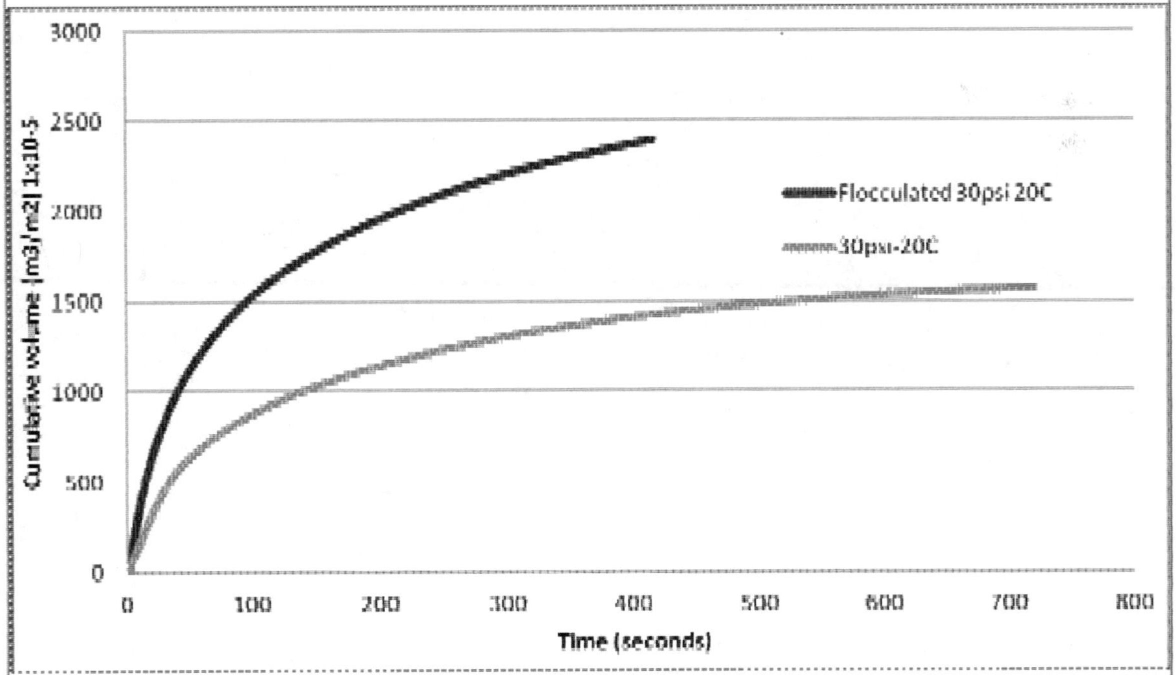

Figure SEQ Figure * ARABIC 3: Effect of addition of PEO

(603f) Lactic Acid Fermentation On Softwood Prehydrolysate by Lactobacillus Delbrueckii

Jing Li, School of Forestry and Wildlife Sciences, Auburn University, Auburn, AL and Maobing Tu, Auburn University, Auburn, AL

Conversion from lignocullulosic biomass into biofuels has been extenstivley investigated recently. However, the whole bioconversion process is still not cost-effective. Developing high-value co-products from biomass prehydrolysate will generate extra revenue to this bioconversion process and become critical to the economic viability of biofuels production. Lactic acid has been classified as valued co-products since it could be used in the food, cosmetic, pharmaceutical chemicals, and biodegradable polymers. The bioconversion process typically includes pretreatment, enzymatic hydrolysis and fermentation. High temperature pretreatment with acid is required to break down the recalcitrant structure of cellulose, hemicellulose and lignin, the harsh condition also generates degraded compounds form carbohydrates and lignin, which are toxic to fermentative microorganisms and inhibit their metabolism, consequently reduces the fermentation rate and yield. The objective of this study is to investigate the fermentability of the softwood prehydrolysate and detoxify the prehydrolysate with different methods to maximize the fermentation yield of lactic acid using *Lactobacillus delbrueckii*. It is has been proved pH control was essential for lactic acid fermentation by *L. delbrueckii* and pH is regulated by adding $CaCO_3$ in this study. Different detoxification methods including alkaline detoxification and H_2O_2 oxidation detoxification has been investigated.

(603g) Production and Properties of Cellulosic Whiskers From Aspen Kraft Pulp

Yang Wang[1], Al Ward[2], Daryl Nichol[2] and Shijie Liu[3], (1)SUNY College of Environmental Science and Forestry, Syracuse, NY, (2)Alberta-Pacific Forest Industries Inc., Boyle, AB, Canada, (3)Paper and Bioprocess Engineering, SUNY College of Environmental Science and Forestry, Syracuse, NY

Cellulose is the most abundant renewable polymer available that can be produced in plants such as woody biomass, cotton, reeds, fruits. A well-known process for preparing cellulosic whiskers from cellulose fibers is acid hydrolysis. In this paper, cellulosic whiskers in the nanometer range were successfully prepared by 16 wt.% sulfuric acid hydrolysis of fully bleached, milled commercial aspen kraft pulp. The effects of preparation conditions, such as hydrolysis time and temperature, on the morphological behavior of cellulosic whiskers were investigated. Higher yields of over 58% were achieved. The chemical composition of the hydrolysis solution was identified and the concentrations of each component were determined by Nuclear Magnetic Resonance (NMR) analysis. The results show the cellulosic whiskers turn the shorter and slimmer when longer hydrolysis time and higher temperature apply. Higher temperature and longer hydrolysis time have negative effects on the yields of cellulosic whiskers, but favor the production of xylose, furfural, glucose, and HMF.

(603h) Identification of Nucleophilic Reactions in Detoxification of Phenolic Model Compounds for Bioethanol Fermentation

Rui Xie and Maobing Tu, Auburn University, Auburn, AL

Biomass pretreatment is typically needed to break down the recalcitrant structure of plant cell wall in bioconversion process. However, during the pretreatment, considerable amount of inhibitors are generated from the degradation of carbohydrates, lignin and extractives; many of these inhibitors significantly reduce the microbial growth and biofuels productivity during the subsequent fermentation process. Overliming detoxification and ammonia conditioning have been widely used for detoxifying biomass hydrolysate. However, gypsum and ammonium salts generated in detoxification will be significant issues for the subsequent processes. Therefore, new detoxification approaches should be investigated for developing cost-effective processes for producing biofuels from lignocellulosic biomass.

In this study, we will use several nucleophiles to detoxify phenolic model compounds and real biomass hydrolysate for bioethanol fermentation with *Saccharomyces cerevisiae*. We have found addition of nucleophiles improved the volumetric ethanol productivity significantly from 0.17 g/L.h to 1.81 g/L.h. This nucleophiles treatment did not change the concentrations of organic acids in hydrolysate, but decreased the furan derivatives by 20%. We will investigate the effects of pH, temperature and reaction time on this nucleophilic reaction. Further research will also be focused on the reaction mechanism between nucleophiles and phenolic model compounds.

(603i) Biomass Characterization Using HPLC and MALDI-MS Techniques

Samira Vasheghani Farahani, Chemical and Environmental Engineering, University of Toledo, Toledo, OH, Michael Mayer, University of Toledo, Toledo, OH, B. Leif Hanson, Instrumentation Center, The University of Toledo, Toledo, OH and Constance Schall, Chemical & Environmental Engineering, University of Toledo, Toledo, OH

Quantitative biomass analyses and molecular weight determination are essential in assessing lignocellulosic feedstocks in biofuel related research. High-performance Liquid Chromatography (HPLC) equipped with a suitable carbohydrate resin column is commonly used for quantitative analysis of polysaccharide constituents in biomass. When sample preparation includes steps such as acid neutralization, introduction of soluble salts can lead to an unstable chromatographic baseline. The contribution of this work is in introducing HPLC sample pretreatment which results in improved baseline stability of resultant chromatograms and therefore reproducible data. Characterizing cellulose, especially its degree of polymerization (DP), is another important step in evaluating its deconstruction to monomeric sugars or in its use for materials. Matrix Assisted Laser Desorption Ionization Time of Flight Mass Spectroscopy (MALDI-TOF) has been used to determine the degree of polymerization for crystalline cellulose with a DP in an excess of 200. A new technique has been developed for MALDI-TOF, utilizing an ionic liquid matrix and solvent system. The technique has potentially wide application in analysis of lignocellulosic substrates.

(603j) Hydrodynamics of a Fluidized Bed Reactor for Fast Pyrolysis of Biomass

Juliana da S. Espindola[1], Fernanda C. Borges[1], Lucas M. Raymundo[1], Luis Antônio R. Muniz[2], Marcelo Farenzena[1], Oscar W. Perez-Lopez[1] and Jorge Otávio Trierweiler[1], (1)Chemical Engineering, Federal University of Rio Grande do Sul, Porto Alegre, Brazil, (2)Universidade de Caxias do Sul, Caxias do Sul, Brazil

Fast pyrolysis is a thermochemical process in which biomass is directly converted into liquid fuels and basic chemicals. Among the technologies applied in fast pyrolysis, the fluidized bed reactors are the most widely used, mainly due to the high mixing performance that this configuration provides, which allows high heat transfer rates and excellent temperature control. In this work the hydrodynamics of a bubbling fluidized bed reactor was investigated, both experimentally and numerically by CFD modeling of fluidizing gas and fluidized particles, in order to allow validation of the model results. The Eulerian approach was used to model the flow of gas and solid phase, both phases treated as interpenetrating continua and governed by a separate set of Navier-Stokes equations. The transport of particles was modeled using drag forces acting on the local volume fraction of solids. The results allow to make a preliminary analysis of the hydrodynamic behavior of a bubbling fluidized bed reactor, helping to understand and develop the process for pyrolysis of biomass.

(603k) Properties and Morphology of Ternary Blends of PLA/PBAT/ Fractionated Lignin

Richard Chen[1], Amar K. Mohanty[2] and Manju Misra[1], (1)School of Engineering and the Department of Plant Agriculture, University of Guelph, Guelph, ON, Canada, (2)Department of Plant Agriculture & School of Engineering, University of Guelph, Guelph, ON, Canada

Lignin is an abundant natural resource that meets every aspect of sustainability since it is highly renewable, low cost, and an industrial by-product of the paper and bioethanol industries. The blending of lignin as a component in a thermoplastic blend poses a challenge in the form of dispersion and compatibility. Polyesters such as poly(lactic acid) and poly(butylene adipate-co-terephthalate) offer the best opportunity of compatibility in melt blending with lignin due to their ability to form hydrogen bonds. The fractionation of lignin into more homogeneous fractions offers better dispersion and more consistent properties. The resulting blend of methanol fractionated lignin, PLA, and PBAT shows highly dispersed lignin particles with particle size of < 5 μm in addition to the presence of lignin between the PLA/PBAT interface inducing improved stress transfer between the two polymer phases. Mechanical properties of the blend showed retention of elongation with slight reduction in strength and modulus for blends containing 30% lignin by weight.

Acknowledgements:

The authors would like to acknowledge the NSERC Biomaterials and Chemicals Research Network for the financial support necessary for this project, along with MeadWestvaco for supplying the lignin used in this project.

(612a) Determination Methods for Caloric Properties of Lignocellulosic Biomass

Karsten Müller[1], Julian Schuster[2], Liudmila Mokrushina[1] and Wolfgang Arlt[1], (1)Chair of Separation Science and Technology, University of Erlangen-Nuremberg, Erlangen, Germany, (2)Chair of separation science and technology, Friedrich-Alexander-University Erlangen-Nuremberg, Erlangen, Germany

To avoid the development and implementation of ineffective utilisation routes for biomass, a systematic process development is crucial. Otherwise, there is a great danger to realize processes that later turn out to have a weak or even negative net energy balance.

A systematic development of new processes requires reliable data concerning the formation properties of the respective biomass. Knowledge of, e.g. the enthalpy of formation, is of vital importance for determining the amount of energy necessary to produce a specific biofuel. Furthermore, these data are essential for the prediction of the reaction equilibrium, which is necessary on account of thermodynamic limitations that may result in decrease in energy efficiency. For some smaller biomass compounds (e.g. certain mono- and disaccharides), experimental data can be found in the literature, but they are virtually not available for most of the higher saccharides like cellulose or complex structures like lignin.

The aim of this project is to examine different methods for the determination of formation properties of long chain and branched polymeric biomass. A possibility for predicting these data are different first- and second-order group contribution methods. Since these models are often not capable to predict all needed properties, yield insufficient accuracy or simply lack in essential parameters, different combinations of technics have been studied and validated.

Short-cut methods for the determination of the enthalpy of formation based on structural information have been developed and tested for their applicability to predict the properties of biomass compounds.

This project has been supported by the Deutsche Forschungsgemeinschaft (Project #: AR 236/34-1).

Abstract #287872

Bark-Based Polyurethane Foams through Solvent Liquefaction

__Ning Yan__, University of Toronto, Toronto, ON, Canada
Jason D'Souza, University of Toronto, Toronto, ON, Canada
Rafael Camargo, Huntsman, The Woodlands, Texas, USA
Steve Diamanti, Huntsman, The Woodlands, Texas, USA

Abstract Text:

Liquefaction of bark is an appealing route to produce polyols for the synthesis of bio-based polyurethane foams (PUFs). Bark is inexpensive since it is a waste material of the logging industry, and its composition is rich in extractive compounds, especially condensed tannins. These phenolic compounds have high hydroxyl functionality, are extracted more easily compared to cellulose or lignin and their aromatic structure can impart thermal stability. However, depending on the liquefaction conditions the polyol structure can vary greatly. These differences are manifested in the foam properties. By characterizing both foam and polyol, a greater understanding of this unique system can be achieved.
Liquefactions reactions were carried out at different temperatures producing polyols with increasing degrees of modification and degradation, respectively. It was found that bark-based polyols can produce polyurethane foams with comparable mechanical properties with commercial standard polyols. However, liquefaction conditions had a significant impact on the polyurethane foam performance.

Title:
Bark-Based Polyurethane Foams through Solvent Liquefaction
Submitter's E-mail Address:
mmisra@uoguelph.ca

Preferred Presentation Format:
Oral

Keywords:
Biological Engineering, Chemical Reactions, Physical Properties and Polymers

Session Selection:
Biobased Materials II: Lignin-Based Materials

First Author
Presenting Author
Ning Yan

Email: ning.yan@utoronto.ca

(612c) Advanced Bio and Nanocomposites From Lignin

Sinto Jacob[1], Richard Chen[2], Xiogang Luo[3], Satvinder Panesar[4], Harekrishna Deka[5], Amar K. Mohanty[6] and Manju Misra[2], (1)Plant Agriculture, University of Guelph, Guelph, ON, Canada, (2)School of Engineering and the Department of Plant Agriculture, University of Guelph, Guelph, ON, Canada, (3)Bioproducts Discovery and Development Centre, Department of Plant Agriculture, Guelph, ON, Canada, (4)School of Engineering, University of Guelph, Guelph, ON, Canada, (5)Centre for Bio-Polymer Science and Technology (CBPST), Elore, Udyogamandal, Kochi, A unit of CIPET (Govt. of India), India, (6)Department of Plant Agriculture & School of Engineering, University of Guelph, Guelph, ON, Canada

Lignin, the second most abundant natural biopolymer in the world, serves as a matrix component for cellulose and hemicelluloses in plant cell walls and provides mechanical strength to biofibres. Currently, about 70 million tons of lignin is generated annually as a co-product in the paper pulp industry . Furthermore, in order to fulfill the demand for lignoncellulosic bioethanol in the United States in the near future, about 225 million tons of lignin generation is expected from the cellulosic bioethanol industry. Only about 2% of the generated lignin is being used for value added applications while the rest is used as burning fuel in the same generating industry. Sustainability of these industries greatly depends upon the value added applications of this co-product. Lignin is an amorphous substance that has potential for material applications. It is a complex polyfunctional macromolecule which is composed of a large number of polar functional groups.

Utilizing the functionality of modified lignin, a series of novel lignin based bioblends/biocomposites have been developed in our laboratory and show tremendous promise for applications in automotives and housing. All of the processed bioblends/bionanocomposites showed higher static and dynamic mechanical properties as compared to the neat matrix containing the same amount of bioresins. This presentation highlights innovative value added non-food industrial applications for underutilized vegetable/plant derived resins and biomass derived biofibers and lignin.

This research is financially supported by the Ontario Ministry of Agriculture, Food and Rural Affairs (OMAFRA)-University of Guelph-OMAFRA (Bioeconomy for Industrial Uses Program 2009 & 2010); OMAFRA -New Directions Research Programs 2009, the Ontario Ministry of Research and Innovation BioCar Initiative project, Natural Sciences and Engineering Research Council of Canada (NSERC)-CRD program; Grain Farmers of Ontario (GFO) fund; and Manitoba Pulse Growers Association (MPGA) fund.

(612d) Lignin - Polypropylene Blends: Formulation and Performance

Hongbo Li[1], **Minh-Tan Ton-That**[2] and Nathalie Legros[1], (1)National Research Council Canada, (2)National Research Council Canada, Boucherville, QC, Canada

Lignin is a polyol and also polyphenol in plant stem and the second most abundant natural polymer coming only next to cellulose. It is an attractive biomass since it is readily available and relatively inexpensive. Currently lignin is mostly used to be burned for energy or to recover for chemicals. Significant research interest has been focused on utilizing lignin as a polymer or filler since last century [1, 2]. However the properties of the lignin based composites by adding lignin as particulate filler to a second polymer are far from satisfactory, due to the complex and heterogeneous structure and chemistry of lignin which are highly dependent on the lignin sources and processing conditions.

In this study, lignin was used to replace partial PP by melt-blending of lignin with PP in a presence of different coupling agents using a twin screw extruder. The effects of lignin concentration and lignin type obtained from different sources and extraction processes on the morphology as well as rheology and mechanical and thermal properties of the lignin-PP blends were systematically investigated. The effect of coupling agent type and concentration on the blend performance was also evaluated. The results obtained indicate that with an appropriate formulation and processing conditions lignin can replace efficiently PP otherwise it can destroy badly the PP's mechanical strength and ductility.

(612e) Physicochemical Properties of Pyrolytic Lignins Derived From Fast Pyrolysis of Etek Lignin

Robert Y. Nsimba, Sustainable Biofuels and Co-Products , USDA-ARS-NAA-ERRC, Wyndmoor, PA and Akwasi A. Boateng, Eastern Regional Research Center, Agricultural Research Service, USDA, Wyndmoor, PA

Producing bio-based chemicals as part of the thermochemical process technologies is becoming increasingly relevant to product diversification envisioned for biofuels and biobased product development. The idea of utilizing the high molecular weight lignin-like oligomers (pyrolytic lignin) in biomass derived fast pyrolysis oil as potential feedstock for the production of renewable chemicals instead of fuels has been discussed. However, the potential of using lignin-rich byproducts in the fast pyrolysis feedstock-to-chemical value chain is yet to be explored. In this study, commercial Etek lignin was subjected to a fluidized bed fast pyrolysis yielding a dark pyrolysate. Following a soxhlet extraction to remove interfering chemicals, subsequent water extraction allowed the fractionation of pyrolysis oil into water-soluble and water-insoluble pyrolytic lignins. Based on the variability in their solubility, pyrolytic lignins were separated into different fractions (ethanol-insoluble, water-insoluble, dichloromethane-insoluble, and dichloromethane-soluble) having different specific molecular weights. The resulting lignin fractions were characterized by elemental analysis, gel permeation chromatography, ^{31}P-NMR, ATR-FTIR, TGA, and scanning electron microscopy. By analyzing the extent of lignin degradation under thermal treatment with the various analytical techniques, we were able to establish the differences in their chemical structure, molecular weight and thermal behavior, and subsequently their potential utility as starting materials for the sustainable production of renewable chemicals and fuels.

Exploring Lignin Based Biorefinery in India

A. K. Ray, Department of paper technology, Indian institute of technology, Roorkee, Saharanpur, India and **Narayan C. Mishra**, Paper Technology, Indian Institute of Technology Roorkee, Roorkee,India, Saharanpur, India

ABSTRACT

Lignin is one of the versatile chemicals , nature has ever produced. These biomaterials are abundantly available in plant biomass and can be utilized as such or as derivatives for various applications. It can be utilized as dispersants for a variety of iorganic materials ,e.g., clay, cement, calcium carbonate, and titanium dioxide, asphalt emulsifiers and emulsion stabilizers, base for industrial detergents(washing powders and liquid soaps),drilling mud thinner, surfactants, coagulants and flocculants(for sewage and waste water treatment), retention aids, fire retarding agents(chlorobrominated lignin),drug formulation, catalyst substrate, carriers for adhesives, insecticides and herbicides, as plastics,binders and adhesives such as lignin reinforced polymers, and graft polymers including polyurethanes, polyesters, polyamines and epoxies, and rubber additives. The main lignin polymers are lignin sulphonates which are used in emulsification, defloccuation, chelation, adsorption and adhesion. The most important applications are for oil well drilling muds, cement and concrete additives, protein precipitants ,sequestering agents, dispersants for a variety of products such as refractories, tiles, absorbants, cosmetics, crayons, medicines, paint, paper coatings, tooth powder and dyes, ore floatation, and electrolytic refining,tanning agents , ion-exchange resins etc. Other uses of lignosulphonates are lime plastering medium, storage battery plates-as expander and component of separator., lime plaster-use of less water with improved settling properties, crystal growth inhibitor, Ingot mold wash, and many more. Lignin- polyisocyanate/polyurethane foam,lignin-urea/phenol formaldehyde, thermosetting resins, lignin-starch, lignin-geletin,glues with polyvinyl alcohol/ acetate, are some of the notable examples as biodegradable polymers and composites. Some low molecular weight compounds like vanillin, dimethyl sulphonates, dimethyl sulphoxide, furfural, methyl mercaptan, phenolics can also be obtained from lignin.

The main source of lignin and its derivatives in pulp and paper industry is black liquor or soda liquor and sulphite waste liquor though lignin can also be obtained from liquor obtained from steam explosion, hydrothermolysis and pure organosolv process.

Presently Indian pulp and paper industry is passing through a crisis from various angles including scarcity of raw materials and has been rated as highly capital , energy and pollution intensive. Majority of mills using non-forest-based raw materials and agricultural residues(rice straw, bagasse, wheat straw etc.) discharge these valuable chemicals to waste on one hand and polluting the environment to a large extent on the other. These chemicals are also non-biodegradable. According to rough estimate for every ton of bleached pulp produced, about 0.5 ton of lignin is either wasted to sewer or used as a low value fuel. Though some Indian industries of the later category have attempted to design recovery plant or bio-methanation process or to recover lignin based chemicals, it is not very wide spread practice. The process of lignin isolation and separation though known from many years, has not been practiced in industrial scale. The major problems lie in the process of isolation of lignin or its substitutes from sulphite liquor as lignin sulphonates or from kraft process ,the thiolignin, and soda lignin from soda liquor by chemical modification.

The profitable utilization of lignin can be possible by various chemical modification process. These include alkylation or arylation(to produce ethers), carboxylation(to form esters), polymeric condensation products with phenols, aldehydes and amines, phenol derivatives with salt of heavy metals, nitration, halogenation(chlorination,chloro-bromination) or oxidation, oxyalkylation,de-methylation,sulphonation or sulpho-methylation to produce lignin sulphonates and many other unit processes.Some of the processes are established in industry but the majority of these are yet to be develop in commercial scale. With the continual increase in price of crude petroleum , there might be a large price margin existing between low cost abundantly available lignin and some high cost petroleum products. The incentive to develop cost effective lignin technology to produce marketable lignin based products replacing the petrochemicals needs to be explored urgently.

In this present paper various alternative schemes for precipitation of lignin, removal by filtration at optimum conditions of temperature and pressure, followed by purification and re-precipitation and finally conversion to value added products are analysed from the view point of commercial applications. These lignin based bio-products are mostly imported. It has been found that some products are quite competitive with other commercial products. But industry is still hesitant to invest funds to manufacture these chemicals from lignin because of fear of high risk. If it is possible to separate lignin economically from spent pulping liquor and to convert it to usable chemicals, the financial returns might be sufficient to support this industry. This is an imperative necessity now for sustainability of nonwood agri-residue based mills in India. In this present investigation a brief account on the lignin utilization in Indian context is reported and a feasibility study on the techno-economics is presented.

INTRODUCTION:

Lignin is an amorphous , three –dimensional highly branched polymer of natural origin., lignin's complex chemical composition and structure , came into sharp focus only recently , after years of research.The complicated structured lignin polymer consists of oxyphenyl propane units , derived from three-substituted cinnamyl alcohols. Their properties differ greatly between conifers, hardwoods and nonwood lignified fibers like wheat straw, rice straw , bagasse , bamboo etc.

1.0 Utilization of black liquor and sulphite waste liquor

Pulp and Paper industry has a very huge lignin potential. According to rough estimate for every ton of bleached pulp produced, about 0.5 ton of lignin is either wasted to sewer or used as a low value fuel. If the means could be found to isolate it from spent pulping liquor and to convert it to usable chemicals, the financial returns might be sufficient to support separation and processing lignin preparations from spent liquor, thus helping abatement of stream-pollution. In many ways Kraft lignin is a versetile raw material. It can be easily isolated by acidification, preferably carried out with carbon dioxide (stack gas), whereby phenolic group are liberated and lignin precipitates. The precipitated lignin can be removed by filtration at elevated temperature and purified further by redissolution in water and reprecipitation by sulphuric acid.

1.1 Lignin Isolation and separation:

Lignin compounds from sulphate black liquor could be precipitated by bubbling CO_2 into it at about 60°C. The precipitated lignin was separated by filtration and purified by dissolving in dry hexane and reprecipitating in ether. Alkali lignin could be isolated, from soda process black liquor by neutralizing it with CO_2 from stack gases to pH 8.3 to 8.5; about 40 % of the lignin was reported[9] to have been precipitated. The mix was heated to 85°C – 90°C to coagulate the lignin, was filtered. Residual lignin could be further precipitated by addition of sulfuric/hydrochloric acid. Isolated lignin was washed first with cold water and then with dilute sulfuric acid and dried on double drum drier.

The slimmy, gelatinous and hard to filter form of the precipitated lignin from black liquor could be changed to a crystalline, easily filterable and quick settling precipitate by mixing small amount of water-immiscible solvent such as chloroform, methelyene bromide, l-nitropropane or the like, with lignin containing liquid such as Kraft/soda black liquor. The chloroform, or other agent, may then be recovered from the filtered liquid by vacuum- or steam distillation. [10]

Chemical modification of Kraft lignin drastically changes its properties and product obtained are of commercial value. Common modifications can be summerised as follows:

1. Formation of ethers by reaction with alkyl/aryl helids
2. Formation of esters by, reaction with carboxylic acid
3. Formation of polymeric condensation product with phenols, aldehydes and. amines
4. Conversion to sparingly soluble phenol derivatives in reaction with various salts of heavy metals
5. Sulphonation to Kraft lignosulphonates
6. Nitration carried out with nitric acid
7. Chlorination with aqueous chlorine
8. Oxidation, with air or oxygen produces derivatives with changed solubilities and melting points
9. Demethylation of lignin can be accomplished by acidic and~ alkaline treatment and results in the formation of cateohol group increasing the reactivity in condensate e.g. with formaldehyde.

1.2 Lignin derivatives[9-11]:

Lignin thus isolated could be converted to large surface area lignins useful in drug formulations for adsorbing acids in the digestive system, due to their extremely high surface area and their acid-alkali resistance. Other suggested uses include carriers for adhesives, catalyst substrate, reinforcement for plastics and carriers for insecticides and herbicides.

A method to produce biodegradable uncoloured lignin solutions was described[12] wherein aqueous lignin solutions such as pulp mill effluent were treated with polyvalent metal cations. The amount of cation should be sufficient to react with lignins, and the aromatics to produce a water insoluble precipitate which was separated from remaining liquor. The precipitate was subsequently acidified with acid solution and was allowed to stand for sufficient time to permit digestion. The precipitate thus formed was separated by centrifugation; and was treated with alkali to produce aqueous basic solution. This solution was found to be effective in controlling Scale formation in boiler water systems, when added to these systems in amount of 5-50 ppm. Thus method not only recovers valuable product from effluent but also reduces the amount of non-biodegradable and toxic materials prior to discharge. In addition colour of the final discharge is also reduced.

Sulpho methylated alkali lignins mixed with aqueous ferrous salt and dichromate and spray dried to a powdered product was reported to be useful as drilling mud thinner.

1.3 Lignin Reinforced Polymers:

A method for forming lignin reinforced polymers was developed by co-precipitating lignin/latex slurry and gas drying the slurry to leave a lignin reinforced polymer free of inorganic salts. In a process, alkali lignin was dissolved in ammonium hydroxides; and the ammonium lignate thus formed was then modified by adding hexamethylenetetramine to the lignate and heating to 90°-$110^{o}C$ for two hours. Thus treated lignin was mixed with latex to give emulsion, which was precipitated by CO_2. When the precipitate was dried at $300^{o}F$ or below, inorganic salts formed in precipitation sublimed leaving lignin/latex polymer.

1.4 Lignin- polyisocyanate/polyurethane foam:

These foams are useful as plastics and adhesives.

Lignin-polyisocyanate foam can be prepared from polyethylene glycol solution of lignin when the later is treated with toluene di-isocyanate at $115^{o}C$ which afforded a foam having good compressive strength and 74% recovery. Polyurethane type foam from lignin preparations and organic isocyanates are also reported.

1.5 Lignin as dispersants:

Method: preparation of ozone treated lignin from Kraft black liquor by treating lignin with oxygen containing 3% ozone at $50^{o}C$ for two hours; and spray drying the resulting solution has been reported. Oxidized lignin was found to be useful as dispersant for a variety of inorganic materials such as Clay, Cement, Calcium Carbonate, Titanium Dioxide.

1.6 Lignin Adducts as surfactants:

Lignin adducts with 3-chloro-2-hydroxypropane-l-sulphonate, chloromethane sulphonate, acrolein and the like compounds are also reported to be useful as surfactants in dyestuff compositions.

1.7 Adhesive from lignin: lignin-phenol-formaldehyde mixtures as adhesives are used for ply wood and other wood based panels

Method: An adhesive from Kraft or Soda black liquor was prepared by reacting the black liquor (dry matter more than 25 %) with a formaldehyde and combining the product with separately prepared phenol-formaldehyde adhesive.

1.8 Lignin as coagulants and flocculants :

Lignin preparations from pulping liquor chemically modified by reaction with compounds such as formaldehyde, triazines, alkylamines were found to be useful as coagulants in treatment of sewage and other wastes. Cationic water soluble lignin amines prepared by reacting Kraft/Soda lignins with the reaction product of epichlorohydrin and a tertiary amine were suitable for a variety of uses such as flocculating agents, coagulants, retention aids, dispersing agents, asphalt emulsifiers and emulsion stabilizers.

1.9 Fire retarding agents:

Chlorobrominated lignins were highly effective as fire retarding agents. In one method, the inflammable material was coated or soaked with an aqueous slurry or alkaline solution of chlorobrominated lignin followed by drying. In addition chlorobrominated lignins could be combined with bonding compositions used for the production of bonded articles from particulate material, such as fibers, chips, granules, saw dust etc. eg. in the formation of chip board, particle board and the like. Chlorobromination was effected by adding chlorine and bromine concurrently or in two step operation. After completion the reaction, the chlorobrominated lignin was filtered, washed and dried.

1.10 Composites

Composites of alkali lignin and biologically active agents such as furgicides, insecticides, nematocides, herbicides and the like are useful as controlled release pesticides. These composites were prepared by intermixing ammonium salt of lignin with active agent under conditions of high shear mixing to form stable emulsion; acidifying the emulsion with mineral acid at pH 4-5 to precipitate the active agent/lignin emulsion with the active agent being physically absorbed within the lignin and drying the precipitate if desired.

1.11 Coke and activated carbon from black liquor

Process for activated carbon by coking or gasification of black liquor: In a process black liquor was heated in absence of free oxygen to an elevated temperature (450°-625°F) and pressure to prevent vaporization of the water, for a period sufficient to produce granular coke. The coke was separated from the aqueous phase to yield an aqueous effluent of low COD. The coke produced might be burned to supply heat, or used as a soil conditioner or carbonized to produce high grade coke for metallurgical purposes or used to produce activated carbon.

Organic values of black liquor have been converted into heat energy and activated carbon by a method which eliminates the recovery boiler from pulping liquor regeneration systems. Black liquor containing 16 % solids or even lower or at any solids content upto 65 % was hydropyrolyzed under very high pressure at 550°-650°F. The organic matter separates as a water-insoluble precipitate along with inorganic constituents such as Na_2CO_3, and minor amounts of Na_2SO_4, Na_2S. The organic matter could be utilized to produce heat energy or activated carbon which itself could be used for purifying effluents from various section of paper mill.

2.0 By-products from lignin

Generally by-products from lignin can be categorized as follows:

1. Low molecular weight compounds
2. Polymeric products

Low molecular weight compounds

The most practical approach to synthesis low molecular weight compounds from lignin is to heat the black liquor for 10-20 mins at 250-290°C together with small amount of alkali and sodium sulphide.

Use of lignin and lignin derivatives: Lignin Sulphonates and Kraft Lignin, Vanillin, Dimethyl sulphonates,

2.1 Vanillin

Vanillin has a wide range of industrial application. Vanillin is the most widely flavouring agent,, commonly known as vanilla, due to its pleasant aroma and its property of enhancing other flavours.. It is universally used as flavouring agent with a property to mask other odours. Vanillin also finds use as a reductant and as material absorbing ultraviolet light. It can be incorporated into various plastics such as regenerated cellulose, polyvinyl ethers and polyamides to protect them against ultraviolet radiation. Ointments for protection against sunburn on the skin can be prepared with vanillin as active agent. Polymeric materials prepared from vanillin and vanillic acid have potential as fibres and films. Generally vanillin is synthesised from Guaiacol or eugenol.

It can be produced from lignin through oxidation of spent liquor with suitable oxidising agent or by extraction by a suitable solvent. There are a few methods for manufacturing Vanillin.

Method I: Catalytic oxidation under pressure in pressure vessel by prior raising pH of black liquor from 9.5 to 13 by sodium hydroxide and then heating for about 2 hours followed by extraction with chloroform.The aqueous layer is acidified with sulphuric acid to bring down the pH to 3.0 and the filtrate is again extracted with chloroform.

Commercial process based on alkaline oxidation of lignin with oxidising agent as cupric oxide(2% wt./wt.), various nitrobenzene compounds, mercuric oxide etc have reported a better yield of vanillin from lignosulphonates. For example cupric hydroxide oxidation of lignin gives 22 % of vanillin at 160°C and 12.5 % at 102°C. In comparison a maximum yield of 23 % is obtained from nitrobenzene oxidation under optimum conditions of 1 hr at 180°C. Table No.2 gives the product obtained by alkaline oxidation of lignin. While vanillin is the most commonly reported simple compound obtained by alkaline oxidation of lignin, there are a number of other products that are formed which may eventually prove to be of commercial interest.

Method II: The black liquor is is dissolved in chloroform and then extracted with saturated sodium bisulphate solution.

Purification of vanillin is an essential step when synthesised from lignin. Most practical way to purify it is to distillate under reduced pressure or co-distillation with saturated hydrocarbon.

2.2 Organo sulphur compounds from lignin

The most important products are DMS and DMSO.

DMSO is used as a catalyst and organic solvent for dyes and other chemical reactions.Since DMSO is an excellent solvent for many carbohydrate materials, it has found application in preparation of sugar derivatives.DMSO also has found utility as a material to protect cellulose tissue from freeze damage. In this case, it has been shown to penetrate-the cell walls better than glycerin and thus prevents excessive concentration of electrolytes in the cell during freezing. It has also been used successfully to protect red blood carpusals against freezing. Widespread interest in DMCO as a solvent for carrying drugs across cell membrane and as a drug itself have been developed.

Depending upon the source of lignin whether softwood or hardwood, the methoxyl contents of Kraft lignin varies from 14-20 %. Since methyl group represents about half of the weight of the methoxyl group, it is theoretically possible to recover 7-10 % of lignin calculated as methyl radicals.

Synthesis of DMS

Method of production of DMS and MM is by heating Kraft spent liquor with sodium sulphide at 200°C and to purify the product obtained. The yields of DMS are about 3 % and MM 0.2-3 per cent on lignin solids.

Commercial process consist of heating concentrated spent liquor with elemental sulphur to a temperature of 200-250°C. The DMS formed is flashed from the reactor along with some water, MM and H_2S and collected by means of a condenser. The crude DMS is scrubbed by passing it through Sodium hydroxide and distilled.

DMS itself does not find any application but is a precursor for its halogenated and oxidised derivatives.

Synthesis of DMSO

This is a compound obtained from lignin with a variety of applications in chemical and pharmaceutical industries. This can be commercially obtained by oxidising DMS by nitrogen-dioxide under optimum conditions. The reaction can be summerised as

$$3Me_2S + 2HNO_3 - 3Me_2SO + 2NO + H_2O$$

The application of DMSO has extended in many fields. It has been used as a outstanding medium for promoting organic reactions and as a reactant solvent itself. The following points outline some of the most potential application of DMSO.

2.3 Polymeric compounds from lignin :

Lignosulphonates:

Bulk of the materials are imported from Russia and China[2-10]

This class of lignin polymers has attracted a great deal of interest because of their diversified uses. They can also be utilised as the base for-industrial detergents such as washing powders and liquid soaps.

2.3.1 Synthesis of lignosulphonates

Lignosulphonates can easily be prepared from purified Kraft lignin by sulphonation with solutions of alkali sulphites at elevated temperature and pressure. Sulphite spent liquor contains lignosulphonates which are formed during pulping reactions. Practical approach to insolate and purify this compound is to remove most sulphite and sulphate by lime addition upto pH of 10.5, more lime is added to the filtrate to give basic calcium lignosulphonate which precipitates in the pH interval of 10.5 - 12.2. More than half the lignosulphonate is thereby removed and separated by filtration. By the addition of sulphuric acid and magnesium and sodium lignosulphonates can be obtained.

2.3.2 Application of Lignosulphonates:

Black liquor /sulphite liquor produces lignosulphones from rice straw, wheat straw, bagasse, bamboo, sarkanda **etc.** and **also from** hardwood. Fundamental studies on the colloidal properties of purified lignosulphonates demonstrated their specific value for emulsification, defloccuation, chelation , adsorption and adhesion uses. It has diversified industrial applications, some major ones are as under:

i.Oil well drilling muds: In the drilling of oil wells, water and dispersants are continuously fed into the hole, to form, together with suspended fine clay, called as 'drilling mud'. Oil in water emulsions in a drilling fluid has been in use since the 1940's. Lignin sulphonates are valuable for stabilizing the oil in water emulsions which give the mud improved properties in control of water loss, reduction of torque on the drill stem, increase of bit life and general improvement of bore hole condition.

ii.Cement and concrete additives: Lignosulphonates are used as grinding aids in Portland cement and concrete. Their function is to reduce the agglomeration of the-ground particles, to reduce surface-tension at the solid-air interference of the particles-and to keep the surface of the grinding media clean & free. It reduces settling time, water reduction andThese can also be used as air entraining agents in concretes. The entrained air gives increased resistance to spalling and cracking of the finished concrete. The dispersive action of lingo-sulphonates is useful in mixing and placing concrete. Incorporation of 0.07 - 0.15 kg of lignosulphonates-concrete admixture per sack of cement will allow water reduction upto 20 %.This results in better compressive strength and durable concrete. In oil well cementing it thus controls settling time and strength.

iii. Dispersants for variety of Chemicals: Lignin sulphonate dispersants usually have a molecular weight in the range of 10,000 - 40,000. The colloidal lignin coats the surface imparting a negative charge to the particle. The similarly charged particles now repel one another and a good dispersion is obtained.This is used as dispersants in refractories,tiles,absorbants,cosmetics,crayons,medicines, paint,paper coatings, tooth powder and dyes. It can also be base for industrial detergents(washing powder and liquid soaps

iv. Ore floatation: Grinding aids,refining of ores as floatation agent, improves filtering properties in Pb-Zn system.

v. Emulsifiers and stabilizers: Lignin emulsifiers are quite competitive with other commercial emulsifier. stabilizing oil-in-water emulsions. Emulsion formed are very stable even under frozen or boiling stage. The stabilization of fire fighting foams that results from the interaction of sodium bicarbonate solution and alum is achieved by an addition of sodium salt of Kraft lignin. Better results can be obtained by using lignosulphonates.

vi. Electrolytic Refining: Lignosulphonates have been successfully used as 'Brightening' agent for the plated surface preventing pitting and deposition of rough surface in electroplating.The electrolytic refining of copper , zinc etc. are some examples.

vii. Binders and adhesives :for pellets ,briquettes and chemicals lignosulphonate enhances the properties of the specific products. Large tonnage of spent liquor solids are used as binders for pellets and briquettes. 2% calcium lignin sulphonate in iron ore gives a pellet ten times as hard as that made with bentonite binder. The most familiar adhesive product is linoleum cement containing principally lignin sulphonates and clay. These cements are also used as adhesive for ceiling tiles, road surfacing,hard durable surface and laying of dust, chemicals such as clay, calcium carbonate and titanium di-oxide resin ingredients,

viii.Resin ingradients: Combination of lignin sulphonates, with low molecular weight urea and phenol formaldehyde resins have received attention as adhesive or binding agent for fibrous materials such as plywood, particle and chip board and paper laminates. Many good quality glues can be produced by incorporating lignin sulphonates with polyvinyl acetate or alcohol.

ix. Rubber additives: As extender, modified reinforcing pigment. Many types of lattices such as **neutral nitrile** and neoprene are compatible with lignin either in the acid or sodium salt form. Mixture of rubber with Kraft lignin are also suitable for use as rubber to glass adhesives.

x. Protein precipitants: Forms insoluble complex with proteins.

xi.Tanning agents: Forms more soluble tanning solution which facilitates recovery procedures.

xii.Sequestering agents: Ca,Ni,Sn,Zn & Al can be sequestered by lignosulphonates above pH 7.

xiii.Miscellaneous uses of lignosulphonates:

Other uses of lignosulphonates are lime plastering medium,stoarage battery plates-as expander and component of separator., lime plaster-use of less water with improved settling properties, crystal growth inhibitor, Ingot mold wash, and many more.

Manufacturers of Lignosulphonates

ABC paper mill, SailaKhhurd, Punjab

15. Status in Research Institutions

Lignin grafted copolymers:Regional Research Laboratory(RRL at Bhubaneswar,Jammu, Jorhat)

Biogas –fuel cell based co-generation:Tata Energy Research Institute (TERI)

Lignin and Its Application with Polymers

Synthetic plastics pose one of the biggest threats to the environment today. The most promising solution to this problem is the prospect of biodegradable polymers. Biodegradable polymers offer the
following advantages. Firstly, they are made from renewable resources and thus do not face the problem of exhaustion. Secondly, they are biodegradable, which implies that the polymer after use will ultimately turn in to compost.. Lignin based polymers present two-fold advantage. One is that they
are abundantly available as tons of lignin is thrown off as a waste product in pulp and paper industries. Secondly, lignin is completelybiodegradable, though slowly. Thus it is
Lignins are highly functionalized biopolymer possessing primarily alkyl-aryl ether linkages, aliphatic and aromatic hydroxyl groups which offer potential for the development of renewable polymeric materials/blends

.

Lignin based polymers and plastics[12-14]:

There are numerous attempts to manufacture entirely New Generation of Lignin-Based Plastics which are biodegradable. Intensive efforts devoted to incorporating lignin preparations into useful polymeric materials have been under way for almost twenty years in a number of laboratories.

A biodegradable plastics have been developed in UM for fabricating 85% underivatized industrial kraft lignin based plastics involves solvent casting (in aqueous 82% pyrrolidine) of blends with polyvinyl acetate and two plasticizers (diethyleneglycol dibenzoate and indene).

Processing and Study of Novel Lignin-Starch and Lignin-Gelatin Biodegradable Polymeric Films[13]

Efforts have been put forward to explore the possibility of extracting lignin from waste wood chips of a paper industry and blending lignin with established biopolymers like starch and gelatin. Lignin is one of the most commonly available natural materials, yet it has been under utilized as a biopolymer. This
paper reports the methodology of preparation and results of tensile and water absorption tests of different lignin- starch, lignin- gelatin and starch-gelatin-lignin polymer films. All the materials used for the polymers are bio derived and completely biodegradable. Solution blending followed by curing was the method adopted for the preparation of the films.

CONCLUSIONS

Lignin in Black Liquors has excellent potential for reuse in Fine Chemical Manufacture. The Paper Industry is totally dependent upon the lignin black liquors as a source of energy fuel in Recovery Boilers.Serious attempts have to be made for medium and small scale manufacturing process.

The chemical and technological aspects of lignin modification and utilization are discussed. Competition with petrochemicals does not, for a number of reasons, permit the commercial utilization of non-oil based products made from lignin. However, pressure from governments and the public for a cleaner environment and new pulping processes (organosolv processes) may inevitably result in full scale industrial use of lignin other than fuel within the not too distant future.

The value addition to the lignin not only improves the economy of the industries but also could serve as a better source for renewable polymeric materials. Research on lignin reinforced thermoplastics and thermosetting polymers composites are under progress. Attempts are being taken to develop low cost, recyclable, fully biodegrdable all green composites from various renewable as well as non-renewable sources that could substitute existing plastics in automotive, building and packaging applications.. Promotion of economy through the material application of lignin, a major industrial byproduct, is the objective of the current research.

REFERENCES

1. Sahoo N, Rohella R.S.,Choudhury S.,Paul S.C.,and Chakravortty, V., Preparation of Value-Added Products From Black Liquor,IPPTA, Vol.8(3), September,1996,pp.45-50.
2. Singh S.P., Recent developments in characterization and utilization of black liquor lignins from small pulp and paper mills.,IPPTA,Vol.2(3),1990,pp.11-15
3. Managaonkar , N.D. and Kulkarni, Y.S.,Preparation and utilization of lignin by-products and lignosulphonates obtained from wood black liquor,IPPTA convention Issue,1987 pp.40-52
4. Ramaswamy ,V. B iotechnology –applications in waste utilization and pollution abatement,ibid, pp.1-13.
5. Jain Sailendra,Environmental protection in Paper Industry,ibid,pp.53-61
6. Dhake, J.D., Utilization of pulp and paper industry waste for chemicals,ibid, pp.19-29
7. Trivedi, M.K., Lignin recovery from alkaline pulping of bagasse, ibid,pp.30-35.
8. NIEM (UNEP) Project Under Phase II- Activity III,October,1990, Institute of Paper Technology(University of Roorkee),Saharanpur-247001(India),pp.64-65,66-67
9. George T. Maloney, ‘Chemicals from Pulp and Wood waste production and application’, Noyes Data Corporation (1978).
10. Agarwal Charu, IIindra ambuj, and Karadbhajne Vijay, Utilization of bBlack Liquor Lignin in Detergent Making, Ippta J.,Vol.23, No.3, July-Sept.2011, pp.181-283
11. J. Johan Lindberg, , Tuula A. Kuusela, and , Kalle Levon1, Specialty Polymers from Lignin, ,*ACS Symposium Series*, Vol. 397
12. Jiju Cherian Vengal, Manu Srikumar, Processing and Study of Novel-Starch and Lignin-Gelatin Polymeric Films, Trends Biomater. Artif. Organs, Vol 18 (2), January 2005 http://www.sbaoi.org
13. R. A. Northey, Low-cost uses of lignin, emerging technology of materials and chemicals from biomass, ACS Symposium Series 476, Washington, D.C. (1992).

(612f) Fully Biodegradable and Biorenewable Ternary Blends From Polylactide, Poly(3-hydroxybutyrate-co-hydroxyvalerate) and Poly(butylene succinate) with Balanced Properties

Kunyu Zhang, Bioproducts Discovery and Development centre (BDDC), Department of Plant Agriculture, Crop Science Building,, University of Guelph, Guelph, ON, Canada, **Amar K. Mohanty**, Department of Plant Agriculture & School of Engineering, University of Guelph, Guelph, ON, Canada and Manju Misra, School of Engineering and the Department of Plant Agriculture, University of Guelph, Guelph, ON, Canada

With the aim of developing novel bioplastic from renewable resource with excellent balance of performances, a fully biodegradable and biorenewable ternary blends from Polylactide, Poly(3-hydroxybutyrate-co-hydroxyvalerate) and Poly(butylene succinate) were firstly fabricated by melt-blending method in this work. The thermal behavior, mechanical properties, and thermal resistance of the blends were investigated. The ternary blends exhibited unique synergistic mechanical properties which are not seen in either of the homopolymers or their binary blends. Mechanical analysis revealed that dispersions of PHBV and PBS minor phases improve remarkably the flexibility and impact strength with reducing the elastic modulus of the PLA matrix slightly. More interestingly, a good balance of stiffness and toughness was attained for PHBV/PLA/PBS ternary blends with PHBV as matrix. The thermal resistance of PLA was improved by the addition of PHBV and PBS. The blends with PHBV as matrix showed more good thermal resistance than that of PLA matrix blends. Enhancement in crystallization of PLA and PBS was achieved by incorporating PHBV phases, whereas the crystallization of PHBV in the blends was suppressed by PBS and PLA. At the same time, perspectives of interesting phase morphology and its relationship to the properties were also discussed.

The authors thank Ontario Ministry of Agriculture, Food and Rural Affairs (OMAFRA) New Directions & Alternative Renewable Fuels Research program (Project SR9235) and New Directions Research Program (Project SR9211) for the financial support to carry out this research work.

(674a) Soluble Inert Template for Preparing Nanostructured Carbon Materials From Forest and Plant Bioproducts

Zhengrong Gu, Agricultural and Biosystems Engineering, South Dakota State University, Brookings, SD and Xiaomin Wang, South Dakota State University, Brookings, SD

Nanostructured carbon materials is critical for multiple purification/separation processes such as water purification, natural gas purification, as well as used as catalyst supports for catalytic chemical processes, and advanced energy storage materials. Soluble inert templates, which reduce preparation cost by simplifying post template removal steps, are most hopeful substitutes for silica or other insoluble porous templates in preparation of nanostructured carbon materials. However, until now, the role of inert salts, such as NaCl, in preparation of nanostructured carbon materials is not clear. In references for ultrasound spray pyrolysis, soluble salts, generated in degradation of precursors or added into precursor solutions, have been recognized as temporary template for synthesis mesoporous or nanostructured carbon materials, while NaCl was identified as chemical activation catalyst for dehydration and irritation in preparing activated carbon with high surface from nuts shell.

In this project, porous structure of carbon materials was also modified by changing salt composition and preparation parameters such as feedstock, temperature procedure, gas composition of atmosphere, and gas flow rate. The pore structure of generated carbon materials were characterized with N_2 isothermal adsorption using a Micromeritics ASAP2010 automated gas adsorption analyzer. Specific surface areas were calculated using the Brunauer–Emmett–Teller (BET) model for the nitrogen isotherm.

Inert soluble salts such as NaCl and KCl were used as templates, while carbon dioxide (CO_2), carbon monoxide (CO), and steam (H_2O) were used as the activating agents to simulate syngas and flue gas process streams in heat treatment of feedstock. The amount of CO_2 in the gas mixture was diluted with inert N_2 while the total flow was held constant to verify catalytic effect of salts. The influence of syngas composition with varying ratios of CO_2:CO is studied to determine the optimum and tolerable amount of CO in the activating agent.

Therefore, nanostructure of carbon materials has been rationally manipulated by changing salt composition and preparation parameters such as feedstock, temperature procedure, gas composition of atmosphere, and gas flow rate in this project. Furthermore, the nanostructured carbon material, with hierarchical porous structure, was prepared from different herbaceous feedstock in this project. The results of this project shown that inert soluble salts, such as NaCl and KCl, not only functioned as template for forming mesoporous structure but also as catalysts for speeding up generation of microporous structure through gasification of biomass in gas containing steam, CO_2 or O_2.

(674b) Alternative Production of Calcium Carbide

Chinedu Umeozor, Chemical Engineering, University of Missouri, Columbia, MO and Tom R. Marrero, chemical Engineering, University of Missouri, Columbia, MO

ALTERNATIVE METHOD OF PRODUCING CALCIUM CARBIDE

Chinedu J. Umeozor and Thomas R. Marrero

Department of Chemical Engineering, University of Missouri, Columbia, MO 65211

ABSTRACT

The primary purpose of this research is to determine the effect of ultrafine reactant particle size on the preparation of calcium carbide. Reactants are lime and biochar, a relatively pure source of carbon. The carbon has a particle size distribution of 10 to 100 micrometers and reacts with lime in a bench-scale furnace. Reaction temperatures are above 1000°C. The anticipated results of these bench-scale studies are (1) the effect of ultrafine particle size on calcium carbide yields as a function of temperature, and (2) determination of the costs/benefits of biochar in the preparation of calcium carbide. These results are significant with respect to calcium carbide quality and the resulting environmental impact from CO_2 emissions for calcium carbide manufacture.

Current CO_2 emissions associated with calcium carbide production are about 1/2 metric ton per ton of calcium carbide. With the global production of calcium carbide exceeding 15 million tons per year, CO_2 emissions to the atmosphere exceed 8 million tons per year. In addition to this issue, for each metric ton of calcium carbide produced, the actual energy consumed is from 3000 to 3800 kWh which differs significantly from the theoretical energy requirement of 2000 kWh. The difference between the actual energy consumed and theoretical energy required to produce a ton of calcium carbide is therefore large and provides an opportunity to reduce the required energy for calcium carbide production and also reduce the associated CO_2 emissions to the atmosphere. This opportunity is addressed in this study by using purer reactants and ultrafine particles. The ultrafine particle size reactants form a eutectic of calcium oxide and carbon which allows calcium carbide manufacture at 500°C less than normal practice.

(674c) Novel Bacterial Isolate for the Conversion of Lignocellulosic Mimics to Bio-Succinate

Jason W. Soares[1], Sanchao Liu[2], Christian Sund[2], Matthew Servinsky[2], Arthur Ragauskas[3], Laurel A. Doherty[1], Cherie Ziemer[4] and Steven Arcidiacono[1], (1)U.S. Army Natick Soldier Research Development and Engineering Center, Natick, MA, (2)U.S. Army Research Laboratory, Adelphi, MD, (3)Georgia Institute of Technology, Atlanta, GA, (4)ARS-USDA, Ames, IA

Bioconversion of lignocellulosic biomass to useful byproducts offers an opportunity to introduce alternative bio-based products into both the commercial and military sectors. Our focus has been on the discovery of microorganisms that exhibit unique lignocellulose hydrolysis behavior while generating useful co-products. A novel isolate, SD CC 1b, has demonstrated hydrolysis of model substrates that mimic the constituents of lignocellulosic biomass, namely cellulose, of varying degrees of crystallinity, and xylan, the major component of hemicellulose. Anaerobic fermentation of SD CC 1b in a mixed gas (85% nitrogen, 10% carbon dioxide and 5% hydrogen) atmosphere, with the respective substrates as the sole carbon source, revealed preferential hydrolysis of the crystalline domains of cellulose relative to amorphous regions after 72 hours. Growth studies on SD CC 1b hydrolysis of xylan, specifically Oat Spelt Xylan (OSX), revealed 6x more growth of SD CC 1b on OSX and 2x more growth on washed OSX (washed to remove monosaccharides present in commercial powders) relative to substrate-deficient cultures after 96 hours. Metabolic analyses predominantly indicated the production of succinic acid, particularly for OSX-containing cultures. During the hydrolysis of OSX, SD CC 1b produced 3 moles of succinate for each mole of acetate while consuming carbon dioxide, which is present in excess in the gaseous environment. Efforts to increase the succinate:acetate ratio by altering the anaerobic atmosphere composition will be discussed. Our novel isolate could contribute to the emerging market of sustainable biomaterials as a renewable, continual source for bio-succinate production through direct bioconversion of lignocellulosic biomass. Bio-succinate could potentially replace petroleum-based succinic acid, which is used in pigments, detergents and as biodegradable polymers in packaging and film applications.

(674d) Polyols and Polyurethane Foams From Crude Glycerol Based Liquefaction of Lignocellulosic Biomass: Effects of Crude Glycerol Impurities On Product Properties

Shengjun Hu, Food, Agricultural, and Biological Engineering, The Ohio State University-OARDC, Wooster, OH, Xiaolan Luo, Department of Food, Agricultural and Biological Engineering, The Ohio State University-OARDC, Wooster, OH and Yebo Li, Department of Food, Agricultural and Biological Engineering, , The Ohio State University-OARDC, Wooster, OH

Crude glycerol, a major byproduct from the biodiesel production process, can be used as an effective and cost-effective biomass liquefaction solvent for the production of polyols and PU foams. However, crude glycerol consists of not only glycerol but also some impurities compounds such as soap and methanol and its exact composition varies widely with different biodiesel production processes. To achieve polyol and PU production of consistent and controllable properties, it is vital to evaluate the effects of crude glycerol impurities on the properties of polyols and PU foams. In this study, the chemical compositions of five crude glycerol samples from different biodiesel plants were first determined, and then the effects of crude glycerol impurities on polyol and PU properties were studied. Our results showed that the composition of different crude glycerol samples varied widely from each other, but all samples were shown to contain glycerol, soap, methanol, fatty acid methyl esters (FAMEs), water, glycerides, free fatty acids (FFAs), and ash as their main components. To investigate the effects of crude glycerol impurities on polyol and PU properties, crude glycerol impurities were mixed with pure glycerol at different formulations and then the mixtures were used as biomass liquefaction solvents for polyol and PU production. Our results showed that the properties of polyols and PU foams were significantly affected by soap, FFAs, and FAMEs: the hydroxyl number of polyols and the density and compressive strength of PU foams decreased significantly with increasing levels of these compounds in liquefaction solvents. Other crude glycerol impurities such as methanol and water were shown to have insignificant effects on the properties of polyols and PU foams.

Sugar Cane refinery -production of sugarcane wax from sulphitation press mud

A.K.Ray , R.B.Nigam* S.K.Sharma*and P.Sanyal

Indian Institute of Technology Roorkee, Saharanpur Campus, Saharanpur , India

*Former Director, National Sugar Institute, Kanpur, Govt. of India

** Former Advisor (By-products),National federation of Cooperation Sugar Factories Ltd.

ABSTRACT

India is the largest cane sugar producing country in the World. Apart from sugar it discharges a large number of byproducts such as bagasse, molasses, press mud , cane tops, leaves and straw. Economy of Indian sugar industry is not as expected due to variety of reasons. Therefore, there are all round attempts all over the cane sugar producing countries to establish a biorefinery to make this industry profitable. One of the waste products of sugar industry is filter mud which is approximately 3.4 % (at 77% moisture) on the basis of 100 tonne cane crushed. The quantity of filter mud varies from country to country, from mill to mill and even within the same mill and from field to field. The parameters influencing the variation include systems of clarification, extraction, cane variety, degree of mechanization etc. Filter mud is not only a source of biological fertilizer, but also helps in the disposal of distillery spent wash after recovering useful fuel in the form of Methane gas. The press mud of sulphitation sugar factories contains a very valuable chemical ,wax which is a suitable substitute of carnauba wax. A number of modified waxes e.g. mountain wax , can also be derived from it. Till today most of the sugarcane industries leave this waste to the field which makes the soil impervious to water penetration if repeated use is made. No attempt has been made to recover wax from the press mud in industrial scale.With ever increasing shortage of vegetable wax, e.g. Carnauba for its multiple uses, these fine chemicals are mostly imported.

If the economy of recovering and refining wax from press mud is established in a sugar refinery , it can be a finest chemical the sugar refinery can produce comparable to the commercialized wax from petroleum refinery on one hand and solving solid waste disposal problem in the sugar industry on the other and considerable foreign exchange can be saved from this exploitation.

In this present investigation ,an attempt has been made to review the potential of various bye products to produce value added chemicals in a sugar refinery with special emphasis on recovery, refining and modification of sugarcane wax . Details of extraction process in laboratory experiments as well as in pilot plant are reviewed. Further investigation is made on economic feasibility studies in order to establish a full sized plant based on data of a pilot plant study.

1.0 INTRODUCTION:

Sugar industry, worth Rs 60,0000 millions is also one of the largest agro-based industries in India-a country that happens to be also the largest consumer of white plantation sugar.[1-2] It has manufactured 26 million MT in 2007-08 which has declined over the following years due to varieties of reasons, 14.7 million tonnes 2008-2009 and again increased to 16.0 million tonnes 2009-2010. With expected 25 million tonnes of sugar production in 2010-12 and 24 million tonnes of consumption, both sugar production and consumption are bound to increase lineally. Indian sugar consumption is growing at a CAGR of 3.75%.Sugar production in India has been on occupies pride of place in the Indian economy being one of the five largest industries comprising of more than 571 units spread all over the country. Sugar industry has many byproducts like bagasse, molasses, press mud, and leaves (trash). Generally for every 100 tons of Cane the products in tons are as follows [1-2]:

Sugar of 98.5o Pol, tons	11.0- 12.5
Cane top	18.4
Leaves and straw	42.9
Final molasses 88^0 Brix	3.0- 3.6
Filter mud 77% -80% moisture	3.0- 3.4
Bagasse 50% moisture	27.5
Flue gases (180°C)	120
Electricity, kWh	1500

Surplus bagasse of 50% moisture content may be of the order of 4 tons.

1.1 Constituents of Sugar Cane:

Sugar cane contains varying amounts of various constituents, approximately 0.2 Per Cent fats and wax, 0.2 PerCent pectins(gums),0.588 percent total nitrogeneous bodies,0.08 Per Cent free acids and 0.12 PerCent combined acids , mostly malic and succinic acid though organic acids present in the cane juice consists of glycollic acid, malic acid, oxalic acids,succinic acid and a small quantity offanoic acid, butyric acid and aconitic acid. These vary from 0.5 to 1.0 per cent of cane by weight. The amount of constituents however depends to a great extent on the variety of cane, its maturity, the condition of the soil, the climate and agricultural practices followed.

Small amount of pentosans along with those constituents find their way into juice during milling process; these together with some coloring matters such as Chlorophyll.,Xanthophyll.,Anthocynin.,Saccharatin,Tannins, polyphenols, nitrogenous compounds such as amino acids(as aspartic.) , albuminoids , asparagin, amides,amino-acids, nuleins,Peptones and xanthin bodies make up the organic solids , not sugar of the juice. Most of the fat and waxes , the greater amount of albuminoids, and part of the gums of the juice are removed during clarification by precipitation. These precipitated impurities ,after removal by filtration, form a cake of varying moisture content called filter muds. This cake contains much of the colloidal organic matter, anions that precipitate during clarification, as well as certain non-sugars occluded in these precipitates. The weight of wet filter muds (80 percent water) averages about 3.4 percent cane. Filter mud contains, on a dry basis, about 1 percent by weight of phosphate (p_2O_5) and about 1 percent of nitrogen. As a result it has been used, especially since the turn of the century, as a fertilizer. The filter mud also contains a mixture of waxy and fatty lipids in a ratio of 5:2 and refined wax can be extracted by appropriate treatment by use of solvents. Sugarcane is also a source of sterols, particularly β-sitosterol, the active ingredient in saw palmetto berries, a nutritional supplement for prostate problems.The active ingredients generally are 80%-98% Octacosanol alone or with Policosanol. Sugarcane is also a source of sterols, particularly β-sitosterol, the active ingredient in saw palmetto berries, a nutritional supplement for prostate problems. The various chemicals can be produced from from sugar industry bye-products. These are depicted in Table 1.

Table.1: Various chemicals from a sugar refinery[1-36]

Raw Sugar /juics	**Molasses**	**Bagasse**	**Filter mud**	**Sucrose**
Ethanol Refined sugar Sugar syrup Dextran Dextranase Epoxy polymer Glycollic acid, Malic acid, Oxalic acids, succinic acid offanoic acid, butyric acid aconitic acid .	Animal feed Ethanol – (Aldehydes &Fusel oil)Rum Vinegar Glycerin Inositol Lactic acid Succinic acid Glutamic acid Aconitic acid Kojic acid Acetic acid, Single cell protein	Xylitol Paper and Paper Board Ethanol Furfural Diacetyl 2,3-pentanedione Composite board(epoxy and bagasse) Compost Animal feed Poultry litter	Fertilizer Animal Feed Compost Waxes and Fats Molasses L-Lysine, Gluconic Acid, Glutamic Acid/Monosodium Glutamate and Itaconic Acid	Methyl *tert*-butyl ether (MTBE) Ethyl *tert*-butyl ether Sucrose syrup Epoxy polymers

1.2 Use of Molasses

(a) Direct utilization: Fertilizer,animal feed
(b) Distilling industry: Rum,ethyl alcohol, rectified spirit,power alcohol,alcohol derivatives
(c) Other fermentation industries: Vinegar and acetic acids, acetone and butanol,citric acid, lactic acid,glycerine, yeast
(d) Other chemical industries: Oxalic acid, and tartaric acid
(e) Miscellaneous industries:Aconitic acid,monosodium glucamate, dextran,L-Lysine, Xanthenes gum,2-keto gluconic,5-keto gluconic , a-ketoglutaric, fumaric acids , glycolic acid,kojic acid and glucaric acid, aldaric acids, Fumaric acid

Molasses contains sucrose. Therefore , a large number of chemicals can be obtained from molasses also from its both sucrose and reducing sugar contents. These are discussed elsewhere[1]

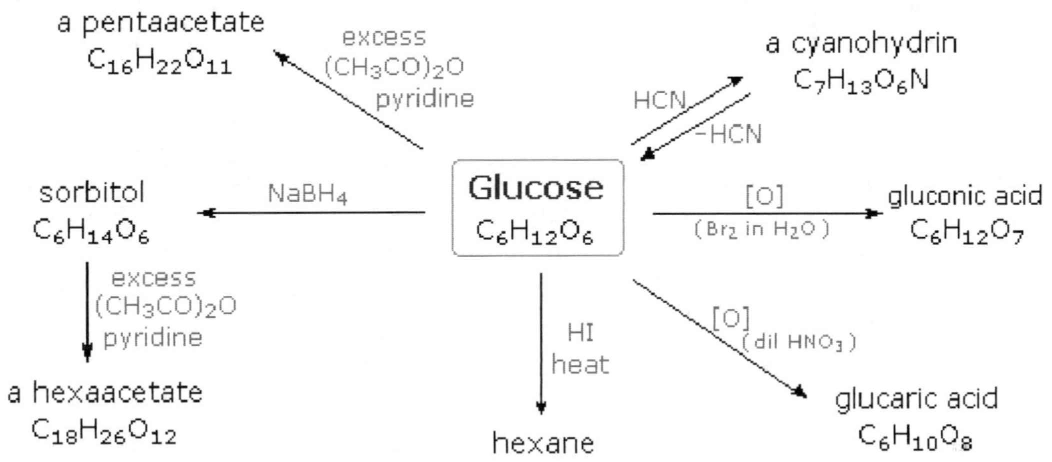

Fig.1 Type of Sugar Industries in India and composition of filter mud:

In sugar Industry in India, generally three types of chemical treatments are accomplished-sulphitation, carbonation and simultaneous sulphitation and carbonation to remove the impurities in cane juice. These impurities are precipitated by the above treatments by using optimum conditions of temperature, and pH, the later after removal by filtration, form a cake of varying moisture content called filter muds. On the above basis , the air dried filter mud will be produced of the order of 5 million tonnes depending upon the moisture content. The weight of wet filter muds (80 percent water) averages about 3.4 percent cane. This mud/ cake contains much of the colloidal organic matter anions that precipitate during clarification, as well as certain non-sugars occluded in these precipitates. Filter mud contains, on a dry basis, about 1 percent by weight of phosphate (P_2O_5) and about 1 percent of nitrogen. As a result it has been used, especially since the turn of the century, as a fertilizer.

The use of filter muds as animal feed has been tried by a number of sugarcane producer territories but so far has not proved economically rewarding, the main constraints being the magnitude of the drying process involved and the low digestibility of the dried scums. The raw material, press mud is a rejected waste material of sugarcane industries, that causes the unwanted problem of pollution to the surrounding suburbs of the sugar mills on accumulation. Also, the availability of the press mud is no problem at all throughout the year. While carnauba plant is a seasonal plant that produces extractable waxes only in marked period of a year.

Press mud of sulphitation sugar factories[16]:

The press mud of sulphitation sugar factories contains a very valuable component in the form of wax. With the increasing shortage of vegetable waxes in the world and new uses coming up for such waxes, the sugarcane wax presents a good source for meeting the future demand for wax not only in India but in other parts of the world.

Carnauba and other vegetable waxes imported into India per annum are valued at a very high price. This shows that a considerable amount of foreign exchange can be saved if sugarcane wax is manufactured in India.

Filter cake or mud press is a chocolate, sticky waste during juice extraction of cane stalks containing 5-15 % protein, calcium and phosphorus; 5-15 % sugar; 10-30% fiber; 10-20% ash and other micronutrients. It is a by-product of sugar industry with high potential as important component of organic fertilizer and animal feed and as source of sugarcane wax and methane (gas for fuel) when process. Sugar cane filter cake mud can also be used in animal feeds. Not only filter mud, but also other supposedly "useless" parts of sugar cane, like molasses and cane tops. Molasses and filter mud in animal feed gives the animals nutrients they otherwise might not be getting, including a carbohydrate (sugar) and B-complex vitamins.

The amount of filter mud varies from country to country, from mill to mill and even within the same mill and from field to field. The multiple factors involved include systems of clarification, extraction, cane variety, degree of mechanization etc. The following table shows the composition of filter mud.

Table 2. Composition of Filter mud[16]

Constituents	A	B	C	D	E
Moisture	72.14	67.73	69.26	69.72	60.331
Sucrose	13.20	10.50	9.96	10.20	0.647
Red. Sugar	0.60	-	-	0.71	0.011
Ash	-	-	7.59	8.10	
Albuminoids	1.93	0.87	1.94	1.80	5.775
Wax	-	3.04	5.13	4.12	7.760

Fibre	-	11.00	-	5.35	10.860
Lime	0.66	0.67	-	1.53	-
Magnesia	-	0.08	-	0.22	0.027
Iron Oxide and Alumina	-	3.06	-	1.40	2.666
Phosphoric Acid	0.57	1.49	1.42	1.39	1.073*
Silica	-	0.21	-	0.37	
Sand & Clay	1.38	1.70	-	2.82	6.506**

*Cal. Phosphate,** Sand and Silica

Sugar cane Wax:

Sugar cane wax is a material that occurs as a thin layer on the outside of the stem of the sugarcane can be obtained as a by-product in the manufacture of cane sugar as a hard green to brown or tan solid.

The filter mud also contains a mixture of waxy and fatty lipids in a ratio of 5:2 and refined wax can be extracted by appropriate treatment by solvents. This wax melts at 82°C, and boils at 146°C, is insoluble in water and in cold alcohol , slightly soluble in cold ether or chloroform, but dissolves easily in hot alcohol, ether, and chloroform, and in benzene [**16**]. This waxy material is not a single body but a mixture of varying proportions of a primary alcohol of the formula $C_{30}H_{62}O$ a with a second body , having the formula $C32H_{68}O$, while possibly yet other alcohols or esters of a high melting point are present in it. It should be noted, however, that only 386 kg of refined wax, which could be roughly equated to carnauba wax, can be obtained from 1000 tonnes of cane. The process is not commercially of interest under existing conditions and, as far as we know, only one plant is operating presently in India and on a small scale.

.How Cane wax goes to the mud?

Cane wax present on the stalk of sugar cane gets partially extracted in to the juice during crushing and gets accumulated in all waste materials , namely juice, bagasse, molasses and the filter mud. In the mud the wax is predominantly higher amount accumulated after defecation and clarification of juice.It is also obtained from vinasses from distilleries and rum factories These solid waste are generally disposed of inlandfills, causing nuisance and pollution due to their high organic and mineral content. The wax from bagasse is of the low order . The yield of wax from sugar cane peels was found to be 0.95% (w/w) of dried sugarcane peels. This yield was less compared to wax extracted from press mud (3.50 to 4.10%) reported by Phukan and Boruah [14].

The percentage of wax in the sulphitation mud is much higher than that in the carbonation mud and therefore makes its recovery from the mud can be partially feasible. The quality of wax present in the filter mud depends on various factors and varies within a wide limit (5 to 21% on dry mud basis).The scale in evaporator typically contains large amount of wax, 40.2 % of the scale in the first evaporator.

Approximate composition of cane wax (carnauba wax) can be as under:

Chemical Composition,%	Values	Fatty Acid,%	Values
Hydrocarbon	51.8	Saturated solid fatty acids	24
Mynicyl alcohol	35.8	Unsaturated liquid acid	30
Saturated fatty acid	12.0	Sterol	42
Sterol	Traces		

Uses of Sugarcane Wax:

Numerous attempts have been made by many researchers [3-35] to use sugarcane wax, after extraction, purification, refining and modification for many applications. .The refined wax has thus to be modified in order to increase its solvent retentivity and gellying properties. Modified cane wax is in every way comparable to carnauba wax. It gives as high a gloss value as carnauba wax and at the same time shows fairly high solvent retentivity.The wax obtained from sugarcane , after refining and modification can be well substituted for the imported materials of the carnauba wax type. Some are illustrated below:

This wax portion finds applications in cosmetics, paper coating, textiles, fruit and vegetable coating, leather sizing, lubricants, adhesives, polishes, and pharmaceutical industry.Thus, it can be considered as an alternative to costly carnauba wax, candelilla wax, and chitosan for edible coating applications. Sreenivas et al.[13] found ash gourd peel a good source of wax which was used as edible coat for strawberries to enhance their shelf life. As the presence of hard wax and vegetable wax is high ,with the industrial expansion of the country; the demand of hard wax is bound to go up. These may be summarized as under[3-15]:

(i) the hard variety sugarcane wax is a replacement of the commonly used carnauba wax. By nature sugarcane wax (the hard variety) is a type of microcrystalline wax having a high potential used in preparation of several important industrial products. Its potential use is wide in industry. The refined variety of the hard sugarcane wax having its potential use in: (a) Medicines and pharmaceuticals (b) foods,in sweet meats and pastries, (c) in tinned food products, (d) in castings and recordings, (e) in carbon papers, (f) in preparation of carbonless copy papers, and carbon papers (g) in electrical insulation of cables and wires, (h) in preparation of various types of paints, varnishes and polishes including boot /shoe polishes, car polishes and floor polishes, (i) in preparation of coloured pencils, crayons, water and oil paints, (j) in cosmetics (k) in preparation of emulsion paints for spraying, ,stencil papers and textile aids, photographic preparations , lubricants and flame proofing agents, corrosion protection, food preservation, tyre plasticizers, printing inks ,toners for copiers and printers, as growth stimulant for seedlings, a source of long chain primary aliphatic alcohols, which find applications as cholesterol-lowering products and other uses. such as protective coatings on fruits for their preservation and for the preparation of waxed papers. Crude sugarcane wax can be used directly for the manufacture of carbon paper

In fact, it has its most likely entries into all the branches of industries where carnauba wax is mostly used. In bulk also it can be produced many times more than any other plant waxes combined together.

(ii) Soft wax portion: It is a mixture of straight chained lipids as established by NMR spectras. It is a source of useful steroids that have their potential and vital uses in medicines and in birth control drugs.

(iii) Pitch: A dark coloured compound of calcium and magnesium salts of wax acids and resin acids. It bears a sticky adhesive character. It has its potential uses in adhesive industries.
(iv) The fibrous residues left over: It forms the biggest part of materials in the constitution of the press mud. The de-waxed residues are enriched with useful organic and inorganic salts like organic matters, nitrogenous salts, phosphates, phosphatides and phosphites, potash bearing elements and salts. Its possible use as a agricultural manure is promising.

With supercritical fluid extraction, the wax is a source of long-chain aliphatic alcohols. Octacosanol (C_{28}) is reported to "increase physical stamina" and "remedy damaged nerve cells" and even to stimulate sex hormones.

Composition of sugarcane wax:

There two varieties of wax available in cane wax, hatd wax and soft wax.

Inarkar and Lele [9June 2012] determined the physical properties of both the varieties. The melting point of crude wax was observed to be 62.1°C. Molecular weight of wax was estimated to be 1706 Dalton. Composition of crude wax found using gas chromatography-mass spectroscopy (GC-MS) was alkanes (28.83%), ester (66.26%), fatty acids (4.58%), aldehyde (0.11%), and alcohol (0.22%). Soft wax portion consists of straight chained lipids as established by NMR spectras. It is a source of useful steroids that have their potential and vital uses in medicines and in birth control drugs

Cane wax from Indian sugar factories:

Press- mud is generally used as a fertilizer in India . Dewaxed mud relations fertilizers properties in a superior manner to the unextracted mud. A survey of wax content in sulphitation press- mud carried out at the National sugar Institute, Kanpur[17,19-36] showed that wax percent dry mud varied from about 20 percent in the various regions of India as given below:

Table 1 Wax content and ratio of Hard wax and soft wax in sulphitation press mud

S. No.	Regions	Average wax content in dry press-mud present,%	Hard wax,%	Soft wax,%
1.	North Bihar	12.3	77.5	22.5
2.	East U.P.	16.1	46.1	53.9
3.	Central U.P.	15.2	69.6	30.4
4.	North U.P.	15.4	66.1	33.9
5.	West U.P.	11.2	70.2	29.8
6.	Bombay	8.7	54.0	46.0
7.	Madhya Bharat	10.0	36.0	64.0
8.	Andhra	20.9	21.8	78.2
9.	Madras	17.3	59.2	40.8

Assuming therefore, that the wax recovery is 10.0 percent on dry mud, there is a potential source in India for the production of **20,000 tons** of sugarcane wax. Sulphitation press mud is at present largely used in India as manure. As the extraction wax from it would improve its manurial value it would be advantageous to extract the wax first before using it as manure.

Extraction of Wax from press mud

The sugar cane wax extraction from press mud and subsequently its modification are very important aspects of by-product exploitation for purposeful economical gain.

For recovering wax, the press mud has to be dried in the sun or by other suitable means and the moisture content is to be reduced to about 3.0 percent or less. The dried press mud is then subjected to a solvent extraction process either of the batch type or of the continuous type similar to the solvent extraction of oil from oil cakes.

Solvents :

Many researches have been carried out on the solvent for wax extraction .Phukan and Boruah [9] used soxhlet extraction of peel powder with carbon tetrachloride (CCl_4) for 8–10 hours. The residue was dissolved in isopropanol and was refluxed with charcoal for 1-2 hours to remove any undesirable color or pigments present. The extract was then filtered to remove charcoal and evaporated at 60°C to obtain crude wax. Inarkar and Lele[9] also used carbon tetrachloride,isopropyl alcohol,toluene. Some other investigators used benzene, cyclohexane at room temperature. For further pextractionof raw wax solvents of reduced polarity such as MeOH, acetone and isooctane were used.

However, for industrial exploitation solvents generally used are (i) n-Hexane (ii) solvent oil B.P. 90—120°C or (iii) mineral turpentine, B.P. 135—190°C. The former two namely (i) hexane and (ii) solvent oil come under the category of 'Dangerous petroleum' and are therefore, subject to strict control by the Inspectorate of Explosives under the Indian Explosives Act. It would be necessary to take all required precautions to avoid explosion, such as the use of explosion- proof motors, gas-proof electrical fittings and switches, storage of solvent in underground tanks, earthing of all steel vessels etc.

Mineral Terpenting

Mineral Terpentine is a safe solvent but requires high pressure steam for solvent recovery inspite of which sometimes the solvent losses are high. The low boiling solvents, however, require apart from safety precautions and special equipment and fittings, suitable solvent recovery equipment involving refrigeration and absorption. The solvent which is to be preferred for the recovery of cane wax should have a boiling range higher than the melting point of wax and in view of the fact that some of the wax fractions have a higher melting point than 80 °C, solvent oil with boiling point range 90°C —120°C is considered the most suitable.

Equipment and Indian Entrepreneur

The simplest equipment would be according to the batch process. The equipment for the batch process would mainly consist of the main extraction vessel with suitable fittings and connections for feeding the mud, feeding the solvent, a perforated false bottom, steam coils for de-solventising the mud, vapour connections to condenser etc.,storage tanks for extracts of different concentrations, a boiling pot for recovering the solvent from the extract with steam connections and vapour connections to condensers, solvent separators, refrigeration condensers, carbon absorption systems etc.

Thus equipment would also consists of solvent pumps with explosion proof motors, wiring through gas - proof conduits and other gas proof electrical fittings. A solvent unit by the batch process has been working at the Ravalgaon Sugar Farm Ltd., Ravalgaon (Distt. Nasik, India) for the last several years and about 400 lbs. of wax is recovered daily by the batch process using mineral turpentine as the solvent. Arrangements have been made to change over to solvent oil in the 90°C-120 °C range but the actual changeover has not taken place.

Various Processes and Operations

The Merz Procss:

While the Ravalgaon plant has been designed and fabricated locally, the Merz plant which also operates on the batch process is as imported plant using n-heptane or n—hexane as the solvent. The usual plant consists of batch type extractors, each extractor having an upper extraction chamber and a lower collecting and evaporating chamber separated by an intermediate head provided with an external tubular heater used to evaporate the solvent, heating and stripping coils, solvent distributor coils, solvent condensers, solvent coolers, water separators etc. The Merz plant is designed to get the fatty acid and hard wax portion from the filter mud separately using the same solvent. The principle on which this plant is based is that the fatty portion is dissolved in the cold petroleum spirit which is separately recovered, after which the same solvent when heated recovers the hard portion of the wax. Each plant consists of an extractor, one fat and one wax still, condensers, cooler, chilled cooler, two automatic water separators, solvent tank, complete refrigeration equipment with chilled water pump etc. The equipment is complete with interconnecting pipes, valves and instruments. The plant can be supplied in various sizes for dealing with 1,2,3 and 5 tons of dry press-mud in 24 hours. The system of extraction has the following advantages:

a. The fatty matter and hard wax can be recovered in the same operation.
b. The same solvent is used for both constituents and solvent losses are negligible.
c. Latent heat of solvent vapour emanating from the **fact** still is utilized for pre- heating of the material in the extractor, thus reducing steam consumption.
d. The working procedure is simple.

One Merz extraction unit for the recovery of cane wax has been installed at Vuyyuru Sugar Factory of M/s K.C.P. Ltd., in Andhra Pradesh (India) several years ago.In this industry ,

Continuous extraction of cane wax.

There are several plants which have boon designed for the continuous solvent extraction of oil cake which with minor modifications can be used for the solvent extraction of cane wax. Manufacturesrs of these units would be in a position to offer suitable plants for these continuous processes. Mention may be made of:

(i) De Smet (ii) Lurgi and (iii) Anderson, Solvent extraction units. The De Smet plant appears to be very suitable for wax recovery from press mud.

The De Smet process:

The De Smet process which is already well known for the extraction of oil from oilseeds may be conveniently applied for the extraction of wax from the press mud. The basic principle of extraction is that of a moving screen upon which the material to be extracted is fed in an adjustable layer. A series of nozzles placed about the belt, spray the solvent on this moving layer of material. The action is countercurrent i.e. the material moves in one direction and the solvent moves in the opposite direction. In this manner the material entering the extractor is first washed by wax containing solvent. Subsequent washes are by solvent which contains less and less wax until the final wash which is fresh solvent from the still. The complete De Smet extraction plant may be divided by function, into three sections.

1. The extractor and its accessory equipment.
2. The miscella distillation unit with stripping column.
3. The meal drying and solvent recovery system.

The De Smet System has the following advantages: Low steam consumption, low solvent consumption, and ease of maintenance, all parts being easily accessible and utmost safety in all operations. De Smet extraction units for extraction of oil are in operation at Kanpur, Calcutta and a few other places in India.

Availability of plant[17]:

The equipment required for the extraction of wax by the batch process can be designed and fabricated locally. The plant would be simple if mineral turpentine is used. If, however, any of the solvents classified as dangerous petroleum are used it would be necessary to have explosive proof motors and gas proof fittings, switches, wiring etc. which may have to be imported. Particulars relating to imported equipment and their suppliers are given below:

Solvent	Plant	Manufacturers	Agent in India
n- heptane	Merz (batch)	Posedown and Thomson Ltd., Old foundary, cannon street, Hull England	M/S Larson and Toubro Ltd. J. K. Building, Dougall Road, Ballard Estate, Bombay, Maharashtra State, India
n- hexane	De Smet (Continuous)	De Smrt S. A. 38, Avenue De France Antworp, Belgium	,,
n- hexane	Anderson (do)	M/S Lurgi West Voltas Germany	,,
n- heptane	Lurgi	,,	IAEC 43, Forbes St. Fort, Bombay- 1

Production of wax from Press mud and the process of refining

The cost of production of crude wax from press mud would depend on. (i) the wax content of mud (ii) the solvent used *(iii)* the type of plant (ii) the capacity of the unit (v) other local factors such as cost and availability of steam, labour etc.
The refining of crude wax, would consist of first de-ashing with concentrated hydrochloric acids,de-fatting it by using suitable solvents such as alcohol, acetone etc., and subsequently bleaching it by chromic or chloric acids according to requirements, cost of chemicals and market for by- products. The bleached and refined wax will have to be modified to suit different end

uses. For polish compositions requiring high solvent retention, and strengths, the wax will have to be modified into the suitable ester and amido derivatives and properly blended.

Scope for developing the wax industry:

As already stated earlier, there is quite a potential for the development of the cane wax industry, for replacing imported waxes, to meet the increasing demand of vegetable waxes in the country and for export, but considerable initial work is necessary before this industry could be established. In the beginning the installed capacity will have to be necessarily small, and there may not be any scope for more than one or two units to be established in the whole country. From the cane wax, a large number of different types of refined and modified waxes for different end uses as has been done by the German manufacrurers by modifying the naturally occurring Montain wax. All the different types of waxes will have to be made and their uses and markets need to be established. In order to establish this industry, the Government might be approached to give the necessary assistance and support for putting up one or two for plants.

Research

A fair amount of research work .has been done in India on various aspects of cane wax extraction refining and modification.

Sugar Cane Wax from Sugarcane Peels:

Wax was extracted from sugarcane peels as reported by Phukan and Boruah [14]. The process involves soxhlet extraction of peel powder with carbon tetrachloride (CCl_4) for 8–10 hours. The solvent was later evaporated using Rota evaporator (Buchi, Switzerland), and the residue primarily contained wax mixtures with some impurities. The residue was dissolved in isopropanol and was refluxed with charcoal for 1-2 hours to remove any undesirable color or pigments present. The extract was then filtered to remove charcoal and evaporated at 60°C to obtain crude wax.
The amounts of Hardwax, softwax and pitch depend on the type of solvents, their ratios and the reactant ratio as well as the type of adsorbents used.

1. Hard wax = 5.20% Softwax= 4.10% Pitch obtained = 0.12%
2. Carbon Tetrachloride+ IPA, Activated Charcoal:Hard wax obtained = 5.35% Soft wax obtained = 4.00% Pitch obtained = 0.12%
3. Toluene+IPA,ActivatedCharcoal:
   Hardwax=5.50%Softwax=4.20%,Pitch=0.15%
   Further,redissolved in carbon tetrachloride,Hard-cream-coloured-wax=5.05%,softwax=3.85%,pitch=0.15%
4. Toluene +IPA, Activated Carbon,: Hard wax = 5.40% (ii) Soft wax = 3.80% (iv) Pitch = 0.12%
5. Benzene+IPA,ActivatedCharcoal : Hard wax = 5.50% Soft wax = 4.20% Pitch = 0.12%

Benzene+IPA,ActivatedCharcoal:Hardwax=5.25%,Softwax=3.85%,Pitch=0.11%

Pilot Plant study:

Das and Rao**[18]** designed and fabricated a continuous fluidized extractor for the exxtractionof sugar cane wax for a capacity of 5 tons of press-mud per day corresponding to a sugar factory of 500 tons cane crushing capacity. The design includes the accessories of pilot plant.

The wax content of press-mud vaies significantly in India. The input material, i.e. crude wax , even if of the order of 1% press- mud on cane it is easy to estimate that with 7.15% of wax in the press- mud a minimum of 96 tons of crude wax for 1000 tons C.T.D. for a working hrs of 120-150 days could be obtained. In India out of 550 sugar factories nearly 460 sulphitation sugar mill will produce minimum45000 tonnes of crude wax in a season.Dewaxed mud retains fertilizing properties superior to the unextracted mud.

The design calculations were based on the following data:

- 20 hrs. per day
- Wax content in press mud=10%
- Ratio of solvent / solid=5:1
- No vapour phase
- Solvent: Petroleum ether-top 80°C-120°C
- Temperature of extraction=75°C

Pilot plant for Modification of Wax:

Nigam[26-35]proposed the design of the pilot plant scale glasslined reaction vessel of 50 gallons capacity to handle 1 ton/day of press mud (dry) for deashing and defatting accompanied by bleaching of wax. The process involving modification of wax involves two steps:

Firstly, deashing is done in which crude wax, water and HCl are mixed by agitation at 90-95 °C for about 2 to 3 hours, it is cooled at room temperature and water separated. Finally a hot water wash is given several times to remove all the acid. Then defatting of the deashed wax is done by thinning the wax with alcohol. Alcohol takes away the fatty portion and granular hardwax is taken out by centrifugation. The alcohol phase is sent to the alcohol recovery unit to recover alcohol and soft wax is used in isolation of sterol. Finally the wax is bleached by chromic acid followed by washings with water to remove all free acid . A part of bleached wax is converted to form ester wax by glycerol and the other to amide wax by ethyl diamine . Then two waxes so obtained can be mixed in proper proportion to get final modified wax.

The vessel is steam jacketed batch type fitted with paddle type agitator which is suitable for deashing and defatting process treatments. As HCl acid is very corrosive acid and that too at an elevated temperature of 90°C,it is suggested to use a glasslined vessel for the purpose.Using experimental heat transfer coefficients available in literature and the natural convective heat transfer coefficients available the steam consumption and heat transfer rates are estimated.

A paddle type of agitator as suggested is a most frequently used type of agitator used for thick paste like this one of wax plus reactants, the scraping action of the agitator dislodges pasty materials from the walls of the vessel . Anchor paddle sweeps the tank bottom also and the entire peripherial area of the tank in contact with the fluid. The standard design variables are selected from the literature for paddle type agitator design and the power required for for it has also been estimated. The schemes for filter-mud bye-products , flow diagram of the solvent extraction of sugarcane wax, refining of wax and modification of wax are shown under:

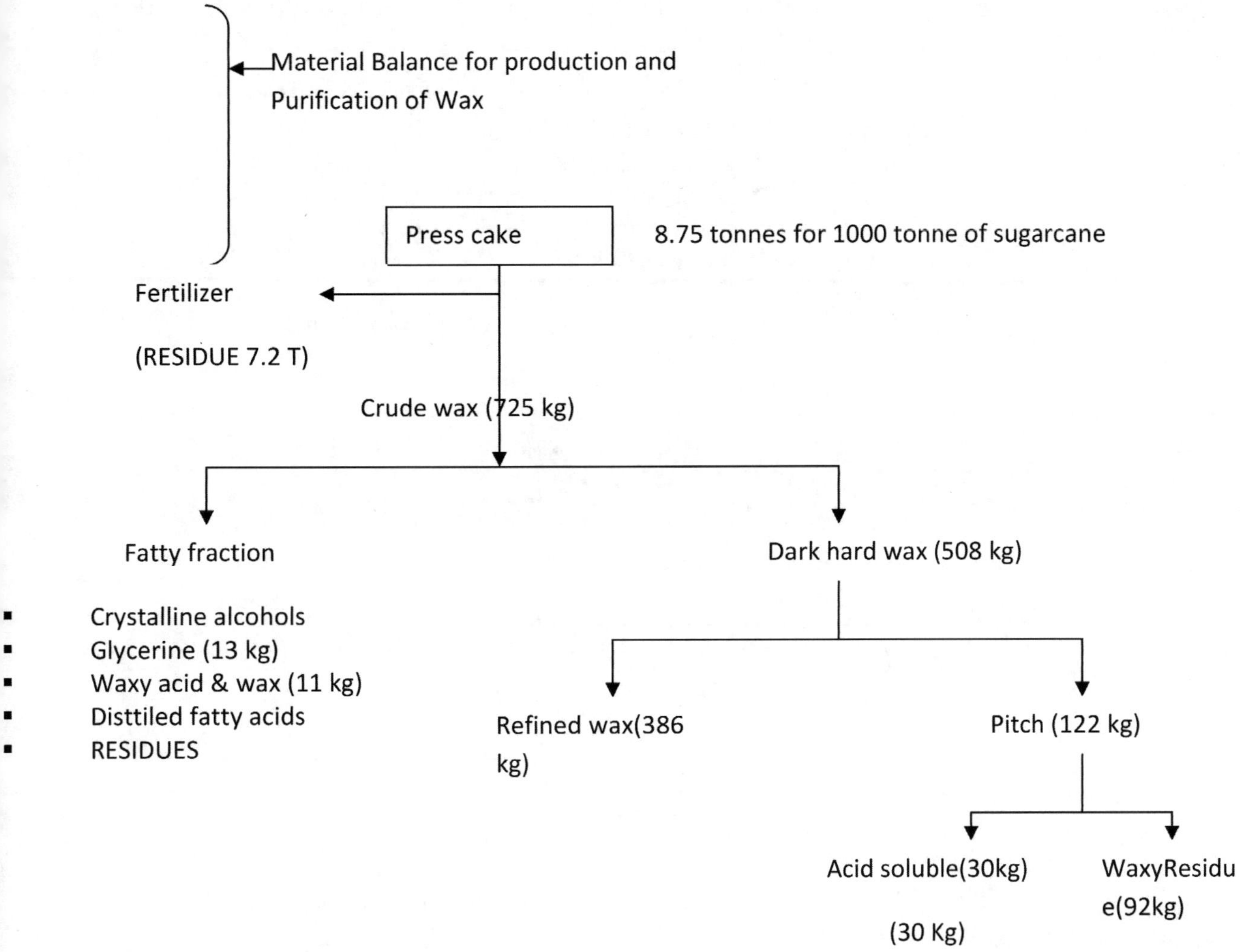

Fig.2 Filter Mud By Products

Flow diagram of the solvent extraction of sugarcane wax

A. Crude Wax Sulphitation factory

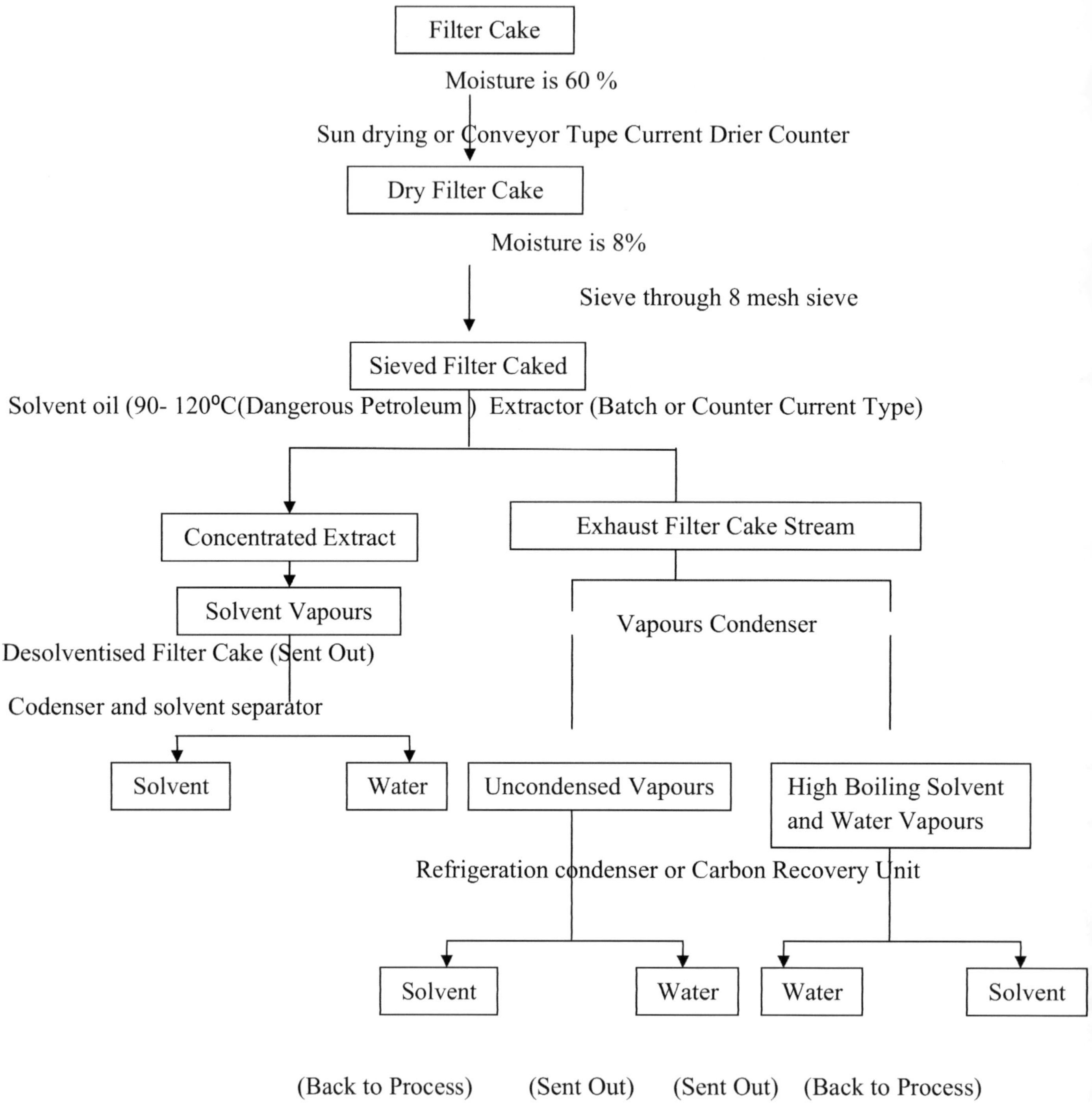

B. Refining of wax

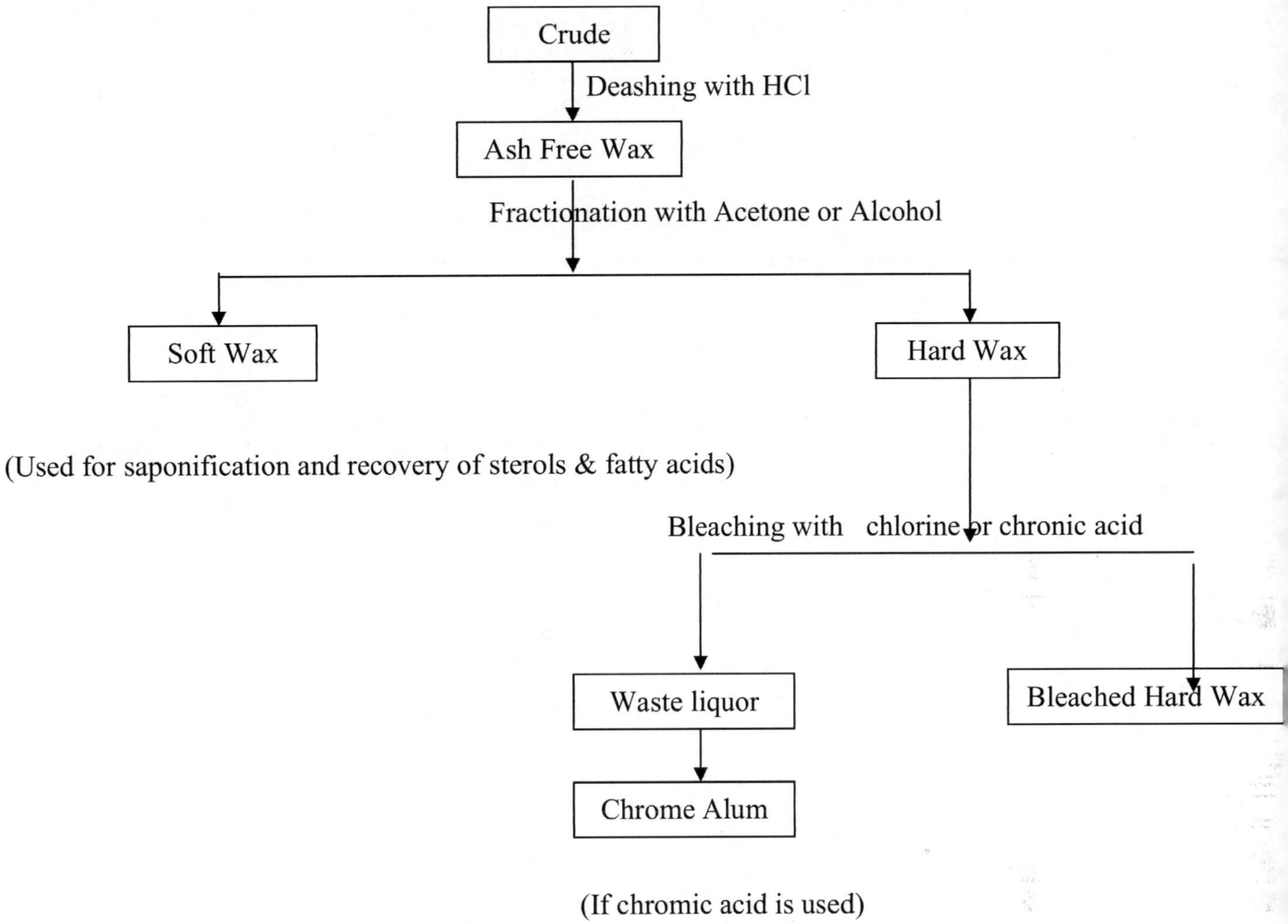

C. Modification of wax

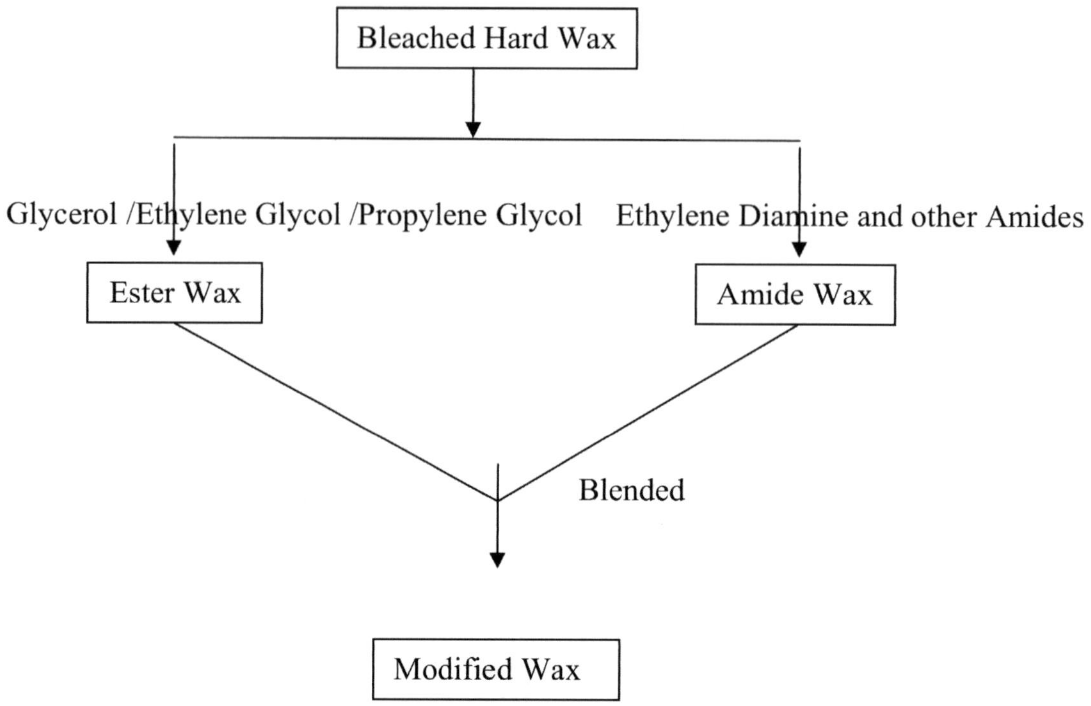

CONCLUSIONS

Extraction of Sugarcane wax has always been a matter of interest to the engineers, due to its enormous industrial applications. But economy of wax recovery has to be studied in detail. Because of these limitations to date, filter mud has not been put to industrial use in a wide scale , although its disposal as a solid waste of the sugar mills is necessary. In most countries, it is spread on the fields. Wax should be recovered from the filter mud which could otherwise make the soil impervious to water if used repeatedly. If properly refined, the wax from filter mud can be similar to the finest vegetable waxed commercialized throughout the world.

Although the chemical possibilities for many bye products from sugarcane seem limitless, few of these products are currently in production. The tradition-bound sugar industry is slow to adopt the new technologies required for implementation. The potential processes require urgent attention. Extraction of Sugar cane wax is one of the processes which has attracted recent attraction.

REFERENCES

1. **A.K.Ray**, Pradosh Sanyal,Alternative Uses of Sugarcane Molasses and Jagerry for production of value added fatty acids ,especially Oxalic acid- Sustainability and Biobased Materials, 17002, Proceedings of 11 AIChE Annual Meeting, held between Oct.16-Oct.21, at Minneapolis 2011,USA,Paper 338e,pp.238

2. **Amiya Kumar Ray**,Pradosh Sanyal,Biorefinery based on Indian Distillery- Innovation of Forest Products, 17000, Proceedings of 11 AIChE Annual Meeting, held between Oct.16-Oct.21, at Minneapolis 2011,USA,Paper 228d,pp.206
3. Sugar Ddyog.com
4. Shrivastava, A. K.Sugarcane wax: physiological significance - extraction and industrial uses, Sugarcane: agro-industrial alternatives. 1995 pp. 333-347, **ISBN** 81-204-0948-5

5. Phukan, Amal C.,Patent 6486335 Issued on November 26,2002
6. Albin H. Warth ,The Chemistry and Technology of Waxes, Reinhold Publishing Corporation,New York,Chapman & Hall Ltd., London
7. http://www.patentstorm.us/patents/6486335/description.html,
8. http://www.ajol.info/index.php/tjpr/article/viewFile/75872/66377
9. Mangesh B. Inarkar and S. S. Lele "Extraction and Characterization of Sugarcane Peel Wax" ISRN Agronomy,Volume 2012 (2012), Article ID 340158, 6 pages doi:10.5402/2012/340158
10. TIFAC,http://www.tifac.org.in/index.php?option=com_content&view=article&id=685&Itemid=205

11. http://bicol.da.gov.ph/Opportunities/sugarcane%20profile/topic11.html
12. A. K. Taylor, "From raw sugar to raw materials," *Chemical Innovation*, vol. 30, no. 11, pp. 45–48, 2000. View at Scopus
13. K. M. Sreenivas, K. Chaudhari, and S. S. Lele, "Ash gourd peel wax: extraction, characterization, and application as an edible coat for fruits," *Food Science and Biotechnology*, vol. 20, no. 2, pp. 383–387, 2011. View at Publisher · View at Google Scholar · View at Scopus

14. A. C. Phukan and R. K. Boruah, "Extraction and evaluation of microcrystalline wax from press mud waste of the sugar industry," *Separation and Purification Technology*, vol. 17, no. 3, pp. 189–194, 1999. View at Publisher · View at Google Scholar · View at Scopus
15. N. Partha and V. Sivasubramanian, "Recovery of chemicals from pressmud—a sugar industry waste," *Indian Chemical Engineering Section A*, vol. 48, no. 3, pp. 160–163, 2006.
16. H.C.Prinsen Geerligs,Cane Sugar and its manufacture,Second Edition,Norman Rodger, 1924
17. Nigam R.B., Unpublished manuscript, National Sugar Institute, 2008
18. Das C.N. and M.N.Rao,Continuous extraction of sugarcane wax(Design of Fluidized Extractor),Journal of Science and Engineering Research, Vol.VI,Part 1, Jan. 1962,pp.127-132,
19. Rao, K.A.N. and Gupta, G. N. Sugar Cane wax from filter press cake, Proc. S.T.A. (ii),79-82,1941,Journal. Sci. and Tech. India VI. 13—15 (1942)
20. Chaturvedi, H.S. ,Gupta G.N. and Sen Gupta, M.N.,Sugar Cane wax from press mud, Indian Sugar, 429-431, 1943.
21. Shrikhande, J. G., A Pilot plant for the extraction of cane wax from press mud, Indian Sugar 73, 1945.
22. Dass, K.S.G. and Kirpa Shankar, Wax Content elimination and subsidar, Scums, J.Sci. and Ind. Res. India,3,462(1945).
23. Ram Dev and Shrikhande, J.G., Chemical Examination of Sugarcane wax from press mud, Proc. STA XV (i), 32—38,1946.
24. Vidyrathi, N.L. and Nansinha Rao, Chemical Investigation of sugarcane Wax ,J. Ind. Chem. Sec. 16, 1939.
25. Ibid, Sugarcane Wax, Indian Sugar, Oct. 1941.
26. Kulkarni, A. B. Indian Patent, 47179 (18th Nov. 1952).
27. Gundu Rao, S.N. Kulkarni, H.G. Kulkarni, D.P., Sugarcane wax, Proc. Pt. I, 62—67 (1949).
28. Kapil, Y.P. and Mukherjee, S., Sugarcane wax and its purification and properties, Part I Proc. STA 1I7 (1952).
29. Kapil Y.P. and Mukherjee, S. Constitution of sugarcane wax composition of Hard Wax, Proc. STA (Part II), 220,228 (1954).
30. Kapil Y.P. and Mukherjee, S., Sugarcane wax— Part II. Purification and properties, Proc. STA (Part II) 229—231 (1954).
31. Mukherjee, S. A survey of cane wax contents of press cake from different sugar factories in India, I.S.S.C.T. Vol.II, 505—512 (1956).
32. Bose, S. and Gupta, K.D., Sugarcane wax — Part III -An examination of fatty lipids for sterols, Proc. STA (Part II), 70—76 (1961).
33. Bose, S. and Shrivastava, A.N. A note on the solvent retentivity characteristics of' sugarcane wax and modified derivatives. Proc. STA. 129—131(1961).
34. Bose, S. and Shrivastava, A.N, Solvent retentivity experiments with sugar Cane Wax and its modified derivatives, Sharkara, 4, 67—72 (1961).
35. Bose,S. and Shrivastava A.N. ,Solvent retentivity experiments with sugarcane wax and its modified derivatives,Part II, Effect of blending different modified derivatives and addition of metallic stearates to the polish composition, Sharkara, 5, 126 (1962).
36. Bose, S., Kapil, Y.P., Mukherjee, S., and Shrivastava,A..N., Purification armd modification of sugarcane wax and study of the characteristics of modified wax derivatives, Proc. I.S.S.C. T., 12th Congress, 1885 (1965).

() 674f: Novel Biodegradable Composites From Poly (Butylene Succinate) Bioplastic and Co-Products From Biofuel Industries: Processing and Properties Evaluation

Andrew Anstey[1], Sudhakar Muniyasamy[1], Murali M. Reddy[2], Amar K. Mohanty[3] and Manju Misra[4], (1)Bioproducts Discovery & Development Centre (BDDC), Department of Plant Agriculture, Crop Science Building,, University of Guelph, Guelph, ON, Canada, (2)Bioproducts Discovery & Development Centre (BDDC), Department of Plant Agriculture, University of Guelph, Guelph, ON, Canada, (3)Department of Plant Agriculture & School of Engineering, University of Guelph, Guelph, ON, Canada, (4)School of Engineering and the Department of Plant Agriculture, University of Guelph, Guelph, ON, Canada

Biocomposites can provide a sustainable platform for biofuel co-products which are available in large quantities due to rapid expansion of biodiesel industries. Co-products function as both a shock absorber and a price adjuster for the biofuel industry by providing an additional income stream. These co-products can be suitable candidates for the development of new biocomposites due to their inherent biodegradability and renewable carbon content.

The present study involves developing biocomposites of poly(butylene succinate) (PBS) using biofuel co-products (canola meal, soy meal, corn gluten meal) as fillers and comparing them with natural fiber-based biocomposites (switchgrass). This work aims to establish the role of these co-products in the design and engineering of new biocomposites, and their influence on material properties and biodegradability under controlled composting conditions. Conventional melt extrusion and injection molding techniques were implemented in fabricating these biocomposites. The mechanical properties (tensile, flexural and impact), thermal properties (TGA, DSC, DMA), structural properties (FT-IR) and microscopic observations (i.e. state of dispersion) (SEM) of these biocomposites were analyzed. A biodegradation study was carried out in order to determine the relative biodegradation rate of the composites and neat bioplastic and to examine the biodegradation process of the materials.

The results demonstrated that the PBS-switchgrass biocomposites had better mechanical behaviour than the meal-based PBS biocomposites. However, the PBS-meal biocomposites displayed better toughness and ductile behavior than the PBS-switchgrass biocomposites. The compostability study revealed that biodegradation rates of the meals were nearly double to that of switchgrass. The same trend was observed in the biocomposites; the PBS-meal biocomposites biodegraded much faster than the PBS matrix and PBS-switchgrass biocomposites. The presence of meals helped the initial growth of microorganisms, leading to hydrolysis of the PBS matrix. This study opens up new knowledge on the comparative compostability of two distinct types of PBS-based biocomposites; one with meal co-products, and another with perennial grass.

Acknowledgements: *This research is financially supported by the OMAFRA – 2009, 2010 New Directions & Alternative Renewable Fuels 'Plus' Research Program-SR9223 and the Ontario Research Fund (ORF) Research Excellence (RE) Round-4 from the Ontario Ministry of Economic Development and Innovations (MEDI) and Hannam Soybean Utilization Fund (HSUF).*

(730a) Investigate Soy Protein As Plastic Component in Polymer Blends

Feng Chen, Washington State University, Richland, WA and Jinwen Zhang, Composite Materials and Engineering Center, college of Engineering and Architecture, Washington State University, Pullman, WA

Investigate Soy Protein as Plastic Component in Polymer Blends

Feng Chen, Jinwen Zhang

School of Mechanical and Materials Engineering & Composite Materials and Engineering Center

Washington State University, Pullman, WA 99164

Abstract

Soy protein (SP) has been recognized as a promising alternative to petroleum-based plastic materials. However, owing to some obstacles such as water/moisture sensitivity, high melt viscosity, narrow process window and/or brittleness, SP plastics based-on neat SP have little practical values. To overcome these disadvantages, blending SP with other biodegradable polymers has received increasing attention. However, SP has only been investigated as organic filler in these blends. Although these blends display higher water resistance, processibility and/or toughness than neat SP plastics, the *stiffness* of the resulting blends are not satisfactory and the *strength* is often lower than that of both the neat SP plastics and the neat polymer.

In this work, soy protein concentrate (SPC) was blended with poly(butylene adipate-*co*-terephthalate) (PBAT) in a single-step compounding process. The mixing of the blends was carried out by extrusion compounding and test specimens were prepared by injection molding. Different influence of SPC as a plastic component and filler on phase structure and mechanical properties of the resulting blends were investigated. The intrinsic cause-deformability of soy protein and extrinsic cause-melt shear stress and interfacial tension in phase morphology development of PBAT/SPC blends were discussed. SP percolation network structure was noted in the blends with SP processed as plastic component. Owing to the extensive crosslinking of the SP molecules, the resulting blends actually became in situ formed polymer/SP composites. The composition-processing-structure-property relationship in soy protein blends were studied in detail. Resulting blends demonstrate an overall superior performance when SPC is used as plastic.

(730b) Urethane Foam Modeling with Soy-Based Polyols

Galen Suppes, Chemical Engineering, University of Missouri – Columbia, Columbia, MO and Yusheng Zhao, University of Missouri, Columbia, MO

Polyurethanes are very important polymers and have many commercial applications. In recent years, polyurethane (PU) foams remain the largest demand sector, accounting for two-thirds of the total PU demand. In addition, this demand for PU foams, especially in construction and transportation, continues to strengthen across the globe.

The polyurethane polymer-forming reaction occurs between an isocyanate and an alcohol. To make foams, the polyurethane polymer must be expanded or blown by the introduction of bubbles and a gas. A convenient source of gas was the carbon dioxide produced from the reaction of an isocyanate group with water. The main purpose of this paper was to initiate Matlab programming to simultaneously model the reacting, volume, and even cell formation during the foaming process.

Modeling is used to advance product development; specifically to expedite the identification of foam formulations where soy-based polyols work well. In these studies, a soy-based polyol prepared by the oligomerization of epoxidized soybean oil with ethylene glycol and a toluene sulfonic acid catalyst.

(730c) Co-Injection Molded Biodegradable Green Composites From Poly (3-hydroxybutyrate-co-3-hydroxyvalerate), Poly (butylene succinate) and Natural Fiber

Vidhya Nagarajan[1,2], Kunyu Zhang[3], **Nima Zarrinbakhsh**[4], Manju Misra[4] and Amar K. Mohanty[5], (1)School of Engineering, University of Guelph, Guelph, ON, Canada, (2)Bioproducts Discovery and Development Centre, Department of Plant Agriculture, University of Guelph, Guelph, ON, Canada, (3)Bioproducts Discovery and Development centre (BDDC), Department of Plant Agriculture, Crop Science Building,, University of Guelph, Guelph, ON, Canada, (4)School of Engineering and the Department of Plant Agriculture, University of Guelph, Guelph, ON, Canada, (5)Department of Plant Agriculture & School of Engineering, University of Guelph, Guelph, ON, Canada

To the best of our knowledge, this work reports the co-injection molding of bioplastics and natural fibers for the first time in the open literature. This approach opens up a novel method of manufacturing green composites for multitude of uses. To strike the stiffness-toughness balance of co-injection molded part, poly (3-hydroxybutyrate-co-3-hydroxyvalerate) (PHBV) and its composites with miscanthus were chosen as the core material while poly (butylene succinate) (PBS) was selected as the suitable skin material. The process of co-injection molding of PBS and PHBV, and also co-injection of PBS and PHBV with miscanthus has been established. The mechanical properties of these co-injection molded parts were compared with those of neat polymers. As expected, the co-injection molded parts' properties were found to be located between the properties of single injection molded skin and core materials. The unnotched impact strength of coinjection molded green composite has drastically increased compared to the PHBV miscanthus injection molded composite. Thickness of the skin and core in the co-injection molded part was measured and flexural modulus was predicted using composite beam theory. SEM micrographs revealed good adhesion between the skin and the core layers. The deformed core confirmed the stress transfer from the core to the skin. These results suggest that co-injection molding technique has been very effective in improving the strength of the composites and the targeted stiffness-toughness balance was achieved.

This research is financially supported by Ontario Ministry of Agriculture, Food and Rural Affairs (OMAFRA)-2008 New Directions research Program Project number SR9211, the OMAFRA, Highly Qualified Personnel Program, Natural Sciences and Engineering Research Council (NSERC) NCE AUTO21 research program and the Ministry of Economic Development and Innovation (MEDI), Ontario Research Fund - Research Excellence Round 4 program.

SYNTHESIS OF VEGETABLE OIL DERIVED POLYURETHANES

Amber R. Tupper, Barbara A. Wheelden, Sunggyu Lee
Ohio University, Athens, OH

Abstract

Bio-based polyurethanes are desired to replace traditional polyurethanes made from petrochemicals in order to be more environmentally friendly and decrease dependence on nonrenewable fossil energy sources. Vegetable oils are a suitable renewable starting material for bio-based polyurethanes due to their availability and low cost. Vegetable oils are triglycerides composed of different fatty acid chains which depend on the plant and the growing conditions. The chemical and physical properties of the oil depend on the fatty acid composition and degree of unsaturation. In this work, polyols are derived from castor oil through epoxidation of the double bonds followed by ring opening of the epoxide group. The thermal properties of the resulting polyurethanes are studied using differential scanning calorimetry (DSC) and thermogravimetric analysis (TGA).

Introduction

Traditionally, polyurethanes are made from petroleum based polyols. The replacement of conventional, petroleum based polymer plastics with bio-based plastics has been a topic of much research and interest over the past decade. The EPA estimates that 31.04 million tons of plastic materials were generated as waste in America in 2010.[1] Roughly 7% of the world's oil and gas consumption goes to the manufacturing of polymers.[2] Polyols from vegetable oils are suitable bio-based alternatives to petroleum polyols.

The process of making polyols and polyurethanes from vegetable oils, including soybean, canola, sunflower, corn and linseed oil, has been the focus of many studies.[3,4] Many studies create polyurethane from unmodified castor oil polyols or castor oil modified by either a transesterfication reaction with another polyol or conversion to an amide.[2,5] One study determined the thermal stability of several polyurethanes from different vegetable oils.[6]

Castor oil was used as it is readily available and has a fairly uniform structure due to the fatty acids composition of up to 90% ricinoleic acid.[5] Due to the hydroxyl group in these ricinoleic acid chains, castor oil is naturally a polyol and polyurethanes can be formed directly from unmodified castor oil. There are about 2.7 evenly distributed hydroxyl groups per triglyceride in castor oil resulting in a uniform crosslinked structure leading to good mechanical properties.[2] Two disadvantages to using unmodified castor oil polyols are that it has a low hydroxyl number and it has a slow cure rate due to the secondary hydroxyls.[7] However, there are other double bonds that can be converted to hydroxyl groups which would result in a higher crosslink density.

Experimental

Materials

Castor oil (Eur. Pharmaceutical Grade), acetic acid (99.8% for analysis), hydrogen peroxide (ACS 30%), methanol (99+%), and fluoroboric acid (50 wt. % in water) were purchased from Acros Organics. Toluene (Histological Grade) was purchased from Fisher Scientific. Amberlite IR-120 (H) (Analytical) was purchased from Supelco. Lewatit MP-64 (Chloride Form) was purchased from Fluka. 4,4'-Methylenebis(phenyl isocyanate) (MDI) (98%) was purchased from Aldrich.

Methods

The procedure to form a polyurethane by the reaction of the polyol, created though the ring opening of epoxidized castor oil, with MDI is similar to that done for other oils and triglycerides.[3,4,8]

Epoxidation of Castor Oil

The epoxidation of castor oil occurs when the oil reacts with peracetic acid which is formed by acetic acid and hydrogen peroxide. Toluene (50ml), castor oil (100g), acetic acid (10g) and Amberlite IR-120 (17g) were added to a 1000 ml 3-neck round bottom flask with a thermometer, drop funnel, condenser and magnetic stirrer. The mixture was heated to 55°C. Hydrogen peroxide was added drop by drop so as to not exceed 60°C and reacted for 7 hours. The Amberlite was filtered off.

Oxirane Ring Opening

The epoxidized castor oil went through oxirane ring opening by reacting with methanol using fluoroboric acid as a catalyst. Methanol (123ml) and fluoroboric acid (260mg) were boiled in a 3-neck round bottom flask. The epoxidized castor oil was added drop by drop into the flask. The mixture was kept boiling for 35 minutes. Lewatit MP-64 was added to neutralize the catalyst and then removed by filtration. The solvent was removed by vacuum distillation followed by high vacuum at 60°C.

Polyurethane Formation

The polyurethane was synthesized by reacting the resulting castor oil based polyols with MDI. The polyols and MDI (158g) were heated to 60°C separately then mixed and poured into molds preheated to 100°C. The polyurethane was cured overnight at 100°C.

DSC analysis

Differential scanning calorimetry was conducted using TA Instruments DSC Q2000 controlled by Qseries Advantage software. The polyurethanes were heated, cooled, and heated to determine the glass transition temperature.

TGA

Thermogravimetric Analysis was done with a TA Instruments SDT Q600 controlled by Qseries Advantage software. Experiments were carried out in an alumina crucible under N_2 atmosphere. Both standard and stepwise isothermal TGA was done.

References

1. Municipal Solid Waste Generation, Recycling, and Disposal in the United States: Facts and Figures for 2010. *United States Environmental Protection Agency*. 2011.

2. Pfister DP, Xia Y, Larock RC. Recent Advances in Vegetable Oil-Based Polyurethanes. *ChemSusChem*. 2011;4:703-717.

3. Zlatanić A, Lava C, Zhang W, Petrović ZS. Effect of Structure on Properties of Polyols and Polyurethanes Based on Different Vegetable Oils. *Journal of Polymer Science Part B: Polymer Physics*. 2004;42:809-819.

4. Zlatanić A, Petrović ZS, Dušek K. Structure and Properties of Triolein-Based Polyurethane Networks. *Biomacromolecules*. 2002;3:1048-1056.

5. Mutlu H, Meier MAR. Castor Oil as a Renewable Resource for the Chemical Industry. *European Journal of Lipid Science and Technology*. 2010;112:10-30.

6. Javni I, Petrović ZS, Guo A, Fuller R. Thermal Stability of Polyurethanes Based on Vegetable Oils. *J Appl Polym Sci*. 2000;77:1723-1734.

7. Ferrer MCC, Babb D, Ryan AJ. Characterisation of Polyurethane Networks Based on Vegetable Derived Polyol. *Polymer*. 2008;49:3279-3287.

8. Petrovic Z, Guo A, Javni I. Process for the Preparation of Vegetable Oil-Based Polyols and Electroninsulating Casting Compounds Created from Vegetable Oil-Based Polyols. . 2000;09/187,992.

Mechanical and Biological Characterization of Corn-Derived Poly-L-Lactic Acid

Leslie Rea[1], Nishath Khan[2], Manju Misra[2], Amar K. Mohanty[3] and Sujata K. Bhatia[4]
(1)Harvard University, Cambridge, MA, (2)School of Engineering and the Department of Plant Agriculture, University of Guelph, Guelph, ON, Canada, (3)Department of Plant Agriculture & School of Engineering, University of Guelph, Guelph, ON, Canada, (4)School of Engineering and Applied Sciences, Harvard University, Cambridge, MA

The ability to create biocompatible materials, and use such materials for medical purposes in the human body, will help treat diseases that result from loss of mechanical function in tissues. Biopolymers are an intuitive choice for biomedical applications such as tissue repair and regeneration, given that bio-based materials are constructed from naturally-derived materials, and may be expected to be friendly to biological tissues. Moreover, bio-based materials possess tunable chemical, physical, and mechanical properties, so that these materials can be readily constructed to match the native properties of a variety of target tissues, and ultimately be implanted in the human body to enable re-growth of cells and tissues.

Congestive heart failure and emphysema are examples of two diseases that result from a loss of elasticity in the muscles and tissue. Both of these diseases are potentially treatable with implantable, naturally-derived biocompatible materials. The purpose of this study is to characterize the mechanical and biological properties of electrospun corn-derived poly-L-lactic acid. We find that corn-derived poly-L-lactic acid has modulus of elasticity of 0.2-0.4 MPa. These properties are compatible with human myocardium, which has a modulus of elasticity of 0.2-0.5 MPa. Further, we find that corn-derived poly-L-lactic acid is non-cytotoxic to clinically relevant cell lines, including cardiac fibroblasts and cardiac myocytes. These results are promising for the use of bio-derived materials in biomedical applications.

Table 1. Processing parameters for electrospinning corn-derived PLLA nanofiber mats.

Parameter	**Sample 1**	**Sample 2**
Wt%	6	7
Weight (g)	6.034	7.2929
Volume (ml)	5	6
Feed Rate (ml/hr)	1.0	1.2
Voltage (kV)	13	15
Distance (cm)	12	12.5
Temperature (^{0}C)	24-25	24
Humidity (%)	63	40-43

Figure 1. Schematic diagram of electrospinning apparatus.

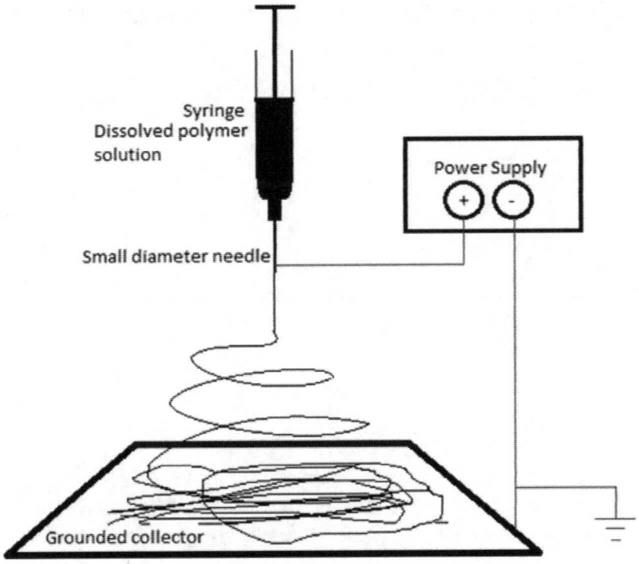

Figure 2. SEM image of an electrospun nanofiber mat constructed from corn-derived PLLA.

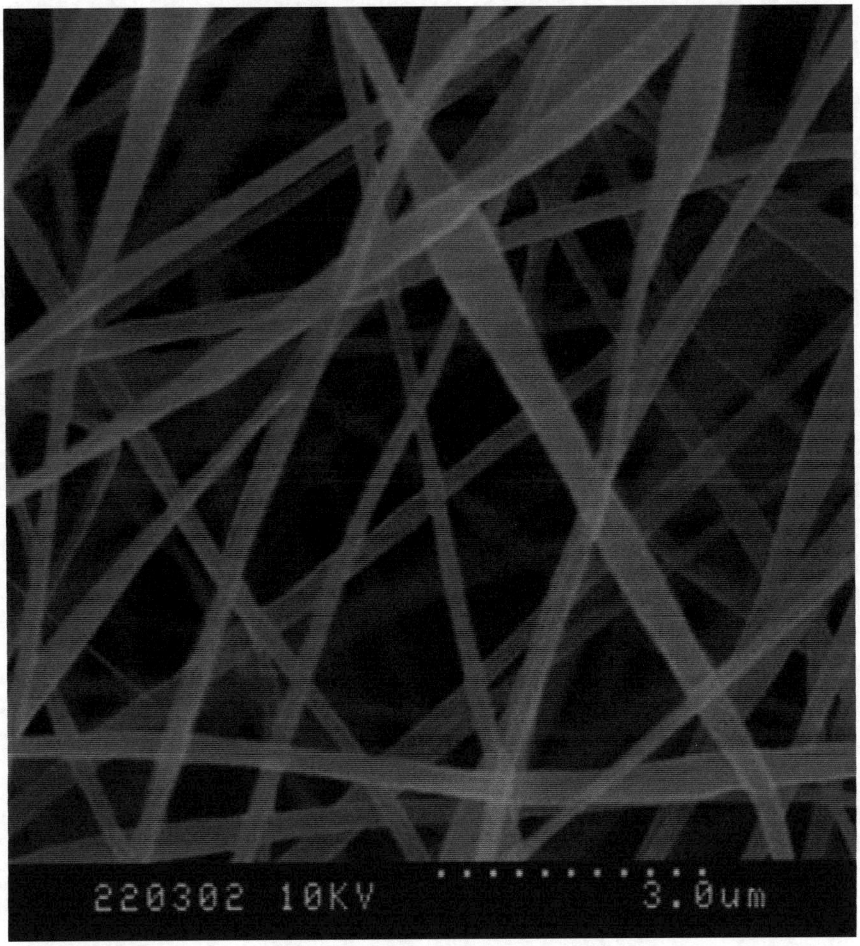

Figure 3. Representative stress–strain plots generated from tensile testing of electrospun nanofiber mats constructed from (A) 6 wt% corn-derived PLLA and (B) 7 wt% corn-derived PLLA.

A

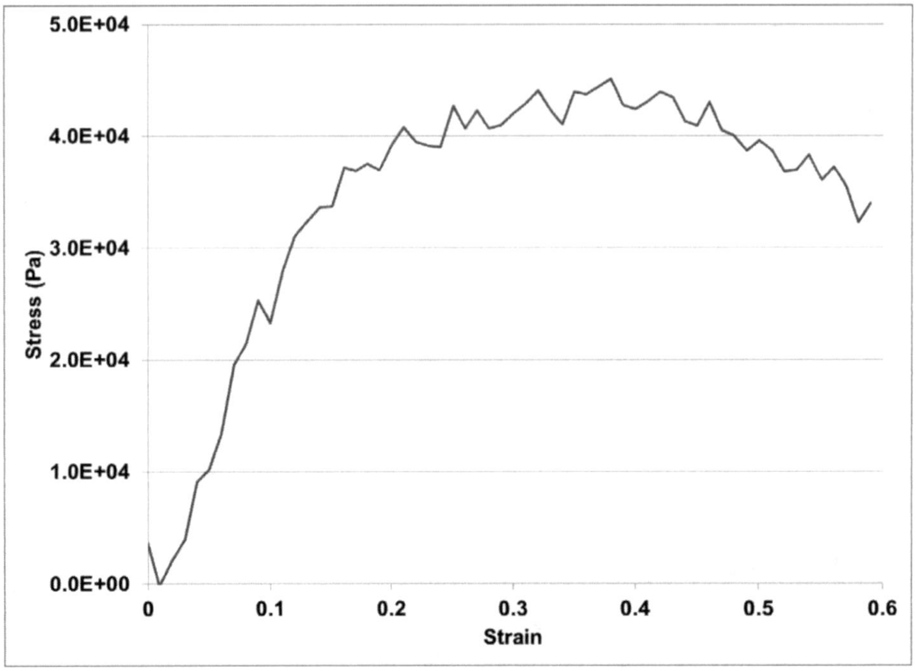

B

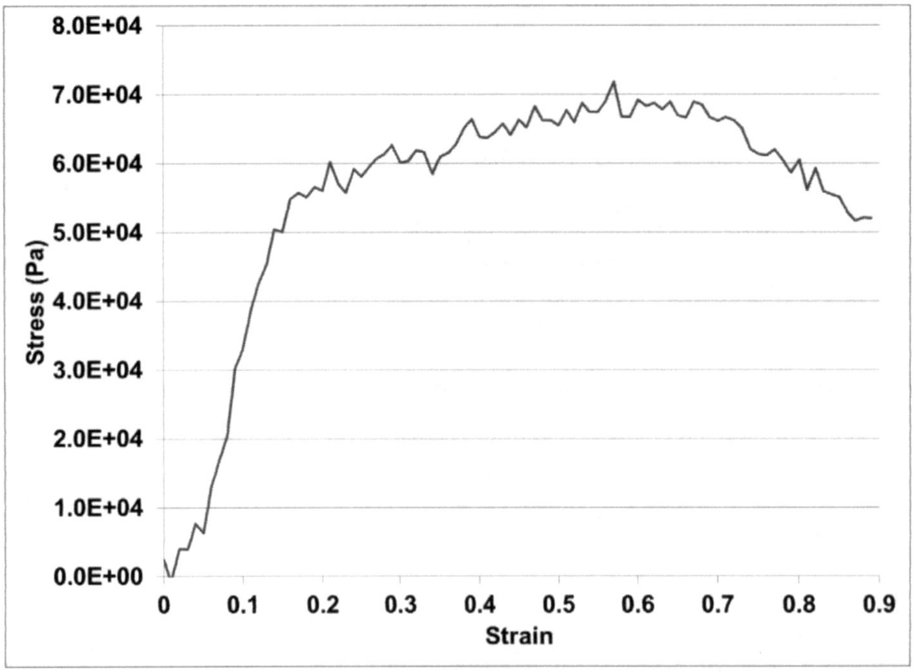

Figure 4. Responses of cardiac myocytes to electrospun nanofiber mats. Myocytes were incubated at 37°C with (A) untreated (control) at 24 h; (B) nanofiber mat constructed from 6 wt% corn-derived PLLA at 24 h; (C) nanofiber mat constructed from 7 wt% corn-derived PLLA at 24 h; (D) untreated (control) at 48 h; (E) nanofiber mat constructed from 6 wt% corn-derived PLLA at 48 h; (F) nanofiber mat constructed from 7 wt% corn-derived PLLA at 48 h.

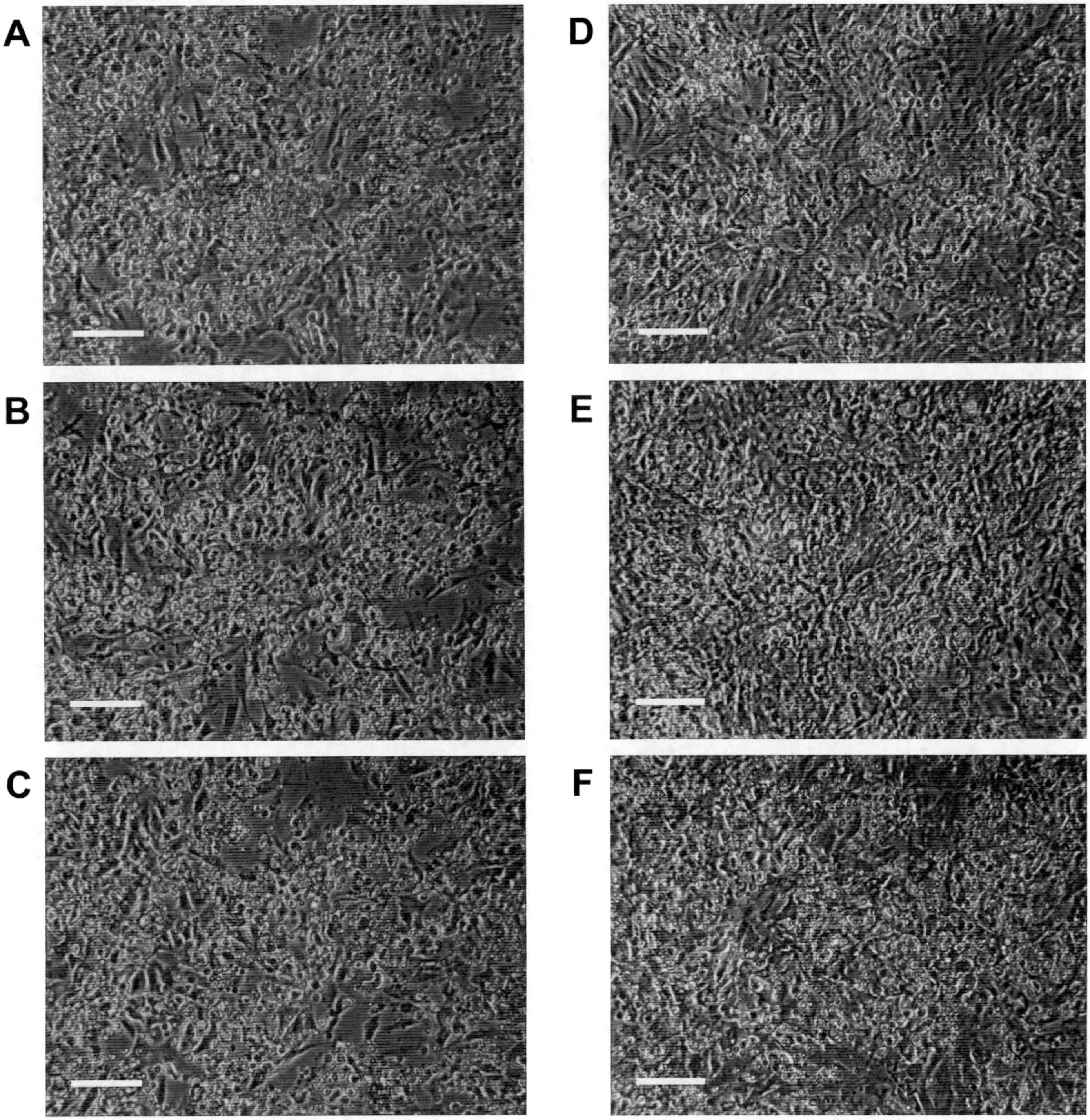

(730f) The Interactions of the Corn Protein Zein and Ionic Liquids

Sean Tomlinson, Jennifer L. Anthony and John R. Schlup, Chemical Engineering, Kansas State University, Manhattan, KS

The corn protein zein is a commercially interesting biopolymer because it is insoluble in water but biodegradable. Ionic liquids are able to dissolve zein and can be used as an extraction solvent for zein recovery. However there has been no research into if or how the zein is affected by the ionic liquid. Using infrared spectroscopy to examine the structure of zein in both protic and aprotic imidazolium-based ionic liquids, we demonstrate how the solvent affects the protein's secondary structure and identify whether the protein is denatured by the ionic liquid. We will also show how the properties of the ionic liquids affect the protein's solubility. It has been found that the ionic liquid's properties are correlated with the structure and solubility of zein in the ionic liquid, particularly the ionic liquid's hydrogen bond accepting and donating abilities. The hydrogen bond accepting and donating abilities of ionic liquids are then used to predict the solubility and the secondary structure of the protein in the ionic liquid.

POLYMER APPLICATIONS IN PULP AND PAPERMAKING- A REVIEW

A.K. Ray*,Sujay Chattopadhyaya and A.K. Mukherjee**
***Department of Paper Technology, Indian Institute of Technology Roorkee**
**** Indian Institute of Technology Delhi**

ABSTRACT

Polymer has numerous applications in the paper manufacturing process either as chemical aid (wet strength and dry strength resins etc. or as a machine components.
In this paper mainly the polymer as wet strength or speciality chemicals are emphasized.
For many practical uses the water and moisture sensitivity of ordinary paper proves to be an insurmountable obstacle. Water leaf paper show excellent dry strength properties, but almost all of this strength is lost when the paper is wetted by water because dry strength is due primarily to hydrogen bonding (and probably some ionic bonding) and these bonds are destroyed by water, which is very effective at forming hydrogen bonds itself. Therefore, development of wet strength requires the formation of different types of bonds which are not adversely affected by exposure to water .The earliest method for developing wet strength was to fuse the cellulose fibers to one another, either by heating the paper to very high temperature or by partially solubilizing the paper with 75% sulfuric acid. The later is the method by which parchment paper is produced. The first chemical method for improving wet strength involved the use of formaldehyde, which can form methylene bridges between hydroxyl groups on the surface of cellulose fibers.

Wet strength is usually reported in terms of the ratio of wet tensile to dry tensile strength, expressed as a percentage. Paper possessing a wet tensile strength greater than 15% of dry tensile is considered to be wet strength papers.

The most important wet-strength agents are applied by beater or wet end addition; i.e., they are added to the pulp before the sheet is formed at the "wet-end" of the machine. Aluminium sulphate is often added to cure resin or to retain anionic Urea-formaldehyde resins. Additives that are not substantive to (absorbed by) paper fibers must be added to the paper after sheet formation. This procedure is usually less cumbersome than addition to the paper stock.

Saturation of paper with water reduces its strength to about 3-10% of its dry strength. Animal glue decreases the rate of absorption of water: if the paper is treated with both formaldehyde and animal glue it will have true wet strength. Treatment with 75% sulfuric acid produces "vegetable parchment," which has high wet strength, but the treatment seriously changes other sheet properties. By means of modern commercial resins, wet-strength of 20- 50% of dry strength can be obtained. Addition of 1% resin can cause a three to six fold increase in wet strength.

INTRODUCTION

Polymers are extensively used in making and manufacturing paper. These are employed either in as chemical aid to increase strengths(wet or dry), as retention aids, drainage aids, de-scaling and flocculating agents , as surface sizing chemicals, coating binders and co-binders,as surfactants and dispersing agents or as a machine parts, paper achine clothing(Nylon,polyester), doctor's blade,roll cover, vacuum boxes,hydrafoil etc. and even in ink for printing[1-10]. In this paper mainly the polymer employed as wet strength or as speciality chemicals are discussed her under:.

WET STRENGTH PAPER:

For many practical uses the water and moisture sensitivity of ordinary paper proves to be an insurmountable obstacle. Water leaf paper show excellent dry strength properties, but almost all of this strength is lost when the paper is wetted by water because dry strength is due primarily to hydrogen bonding (and probably some ionic bonding) and these bonds are destroyed by water, which is very effective at forming hydrogen bonds itself. Therefore, development of wet strength requires the formation of different types of bonds which are not adversely affected by exposure to water .The earliest method for developing wet strength was to fuse the cellulose fibers to one another, either by heating the paper to very high temperature or by partially solubilizing the paper with 75% sulfuric acid. The later is the method by which parchment paper is produced. The first chemical method for improving wet strength involved the use of formaldehyde, which can form methylene bridges between hydroxyl groups on the surface of cellulose fibers.

Wet strength is usually reported in terms of the ratio of wet tensile to dry tensile strength, expressed as a percentage. Paper possessing a wet tensile strength greater than 15% of dry tensile is considered to be wet strength papers.

The most important wet-strength agents are applied by beater or wet end addition; i.e., they are added to the pulp before the sheet is formed at the "wet-end" of the machine. Aluminium sulphate is often added to cure resin or to retain anionic Urea-formaldehyde resins. Additives that are not substantive to (absorbed by) paper fibers must be added to the paper after sheet formation. This procedure is usually less cumbersome than addition to the paper stock.

Saturation of paper with water reduces its strength to about 3-10% of its dry strength. Animal glue decreases the rate of absorption of water: if the paper is treated with both formaldehyde and animal glue it will have true wet strength. Treatment with 75% sulfuric acid produces "vegetable parchment," which has high wet strength, but the treatment seriously changes other sheet properties. By means of modern commercial resins, wet-strength of 20- 50% of dry strength can be obtained. Addition of 1% resin can cause a three to six fold increase in wet strength.

A. **USES OF WET STRENGTH PAPERS**

Wet strength papers are widely used for packing purpose, for making multi-wall paper sacks, paper bags and wrapping papers and containers employed for protection of flowers and moist or frozen food, such as, fish, vegetable and ice-creams, towels, paper handkerchiefs and toilet tissues. Wet strength paper is now frequently used for horticultural labels, currency notes, laundry tags, maps, blue prints, posters and other articles subject to out-of-doors exposure.

Various Wet strength agents and their Curing Conditions.

Type of Resin	Type of Curing
Urea-Formaldehyde (UF)	Acid
Melamine -Formaldehyde (MF)	Acid
Polyamine and Polyamide : Polyamide Epichlorohydrin(PAE)/Polyamine epichloro alkaline hydrazine	Acid
Glyoxal	Acid
Dialdehyde starch (DAS)	Acid
Chitosan	Alkaline
Miscellaneous(neoprene,Styrene–Butadiene,Acrylic butadiene arynitrile latex	Alkaline
Newer Resin (Polyethylenimine, Polyacrolein,	Alkaline
Quarternary Ammonium Epoxide	
Polyacrylamide-Glyoxal (PAMG)	Acid
Quarternary Ammonium Epoxide	Alkaline

B. 1. **UREA -FORMALDEHYDE RESINS (U.F.)**

Nonionic, anionic and cationic urea formaldehyde resins are used as wet strength resins for the manufacture of paper.

a)Nonionic resins are unsuitable for beater addition because they are not retained by pulp. These resins, therefore, must be applied by surface treatment of the paper after formation on the machine. Surface application of urea-formaldehyde resins yields a stiff paper with reduced absorbency of water surface application and is used where properties, such as, wet and dry abrasion resistance, wet-rub resistance, and improved creep are desired.

The bulk of wet strength paper is produced by wet-end addition because of these advantages; it is not necessary to re-dry the paper after treatment of the web with wet-strength agent, no special application equipment is required; uniform treatment is more easily achieved; folding endurance is increased (surface application of U-F resin reduces folding endurance); tearing strength is reduced by surface treatment. Furthermore, wet-strength resins together with aluminium sulphate promote retention of ‘fines’ (short cellulose fibers or fragments) improve the ‘formation’ (uniformity of distribution of fibers in the sheet) and decrease fiber wastes.

b)Anionic Resin: Reaction of urea-formaldehyde polymer with sodium bisulfite produces an anionic bisulfite resin.

```
---------N---------------CH2  -  N----------------CH2  -  N  -  CH2---------------------
         l                       l                        l
         C=0                     C=0                      C=0
         l                       l                        l
         NH                      NH                       NH
         l                        l  -  +                  l
         CH2OH                   CH2SO3 Na                CH2OH
```

The product is water-soluble, but addition of aluminium sulfate provides good retention of material by unbleached kraft-pulp. Paper pulp itself is anionic, and the aluminium sulphate functions as a cationic mordant in approximately the same way that it anchors resin size and anionic dye to paper pulp. Although the anionic U-F resin provides a low-cost means of imparting wet -strength to unbleached kraft paper, they are not especially effective with other pulps and are absorbed somewhat slowly. Its retention with the pulp when added to aqueous pulp suspension is about 40-50%.

Anionic U-F resin is made by reacting urea and formaldehyde in the molar ratio 1:1.8 - 2.0. The pH is made neutral and to it added 11 gm sodium Meta bi-sulphite per mole of urea. The mixture is heated to boiling point, under reflux. After boiling for 30 minutes and boiling must be continued and then the solution is acidified to pH 2.5 with dil. H_2SO_4 for further one hour. The solution is then neutralized with aqueous sodium hydroxide and cooled.

c) Cationic resin is preferred for other than bleached Kraft pulps. They owe their cationic nature to amine groups, which are introduced by reacting ethylene diamine, diethylene-triamine, or other amino compounds with urea-formaldehyde resins. At low pH, the amino group picks up protons, and the resin becomes cationic. In addition to making the resin cationic, the amine increases the solubility of the polymer permitting the polymerization to be carried to higher molecular weight.

The resin is prepared by reacting urea and formaldehyde under mild alkaline condition to give methylol urea, adding the base, or a suitable salt of the base and excess acid and then condensing at elevated temperature under acid conditions until the solution has attained an appropriate viscosity. The syrup is then neutralized and cooled.

Cationic U-F resins available commercially are to be diluted with water before use to ensure a better distribution of resin in fiber. Sufficient alum should be added to the stock to ensure maximum retention of resin in the paper. The acidity of the stock should also be adjusted to pH 4-5 by adding alum and hydrochloric acid. The quantity of resin should not be more than 3-4% of the dry weight of fiber. 60-70% of the resin added to the aqueous pulp suspension is normally retained on the fiber.

Urea-formaldehyde resins generally react less rapidly than melamine-formaldehyde resins, which are, in general, lower than epichlorhydrine-amine resins. The use of a rapidly curing resin permits the mill to sell the paper as soon as it is manufactured, without storage and warehousing. However, repulping of "broke" (trimmings and rejected papers produced in the mill) is more difficult if wet-strength resin cures too rapidly. A cationic U-F resin requires ten to fourteen days to develop full cure (at 20^{0}C) whereas the anionic resin requires about twice as long. Paper containing U-F resin can be repulped.

Urea formaldehyde resins do not show great resistance to hydrolysis. At a pH of 4-6, these resins can be hydrolyzed fairly readily, but they resist hydrolysis at pH 8-10. It has been reported, however, that under favorable conditions, paper treated with U-F resins will retain its full wet-strength for six to eight years.

Characteristics

- Low viscosity liquid Or Powder
- Solid content 68 ± 2(%)
- Shelf life (20^{0}C):

Liquid: 3-6 months
Powder: 1 year
Pot life (20^{0}C) :0-48 hrs.
At usual room temperature:Setting time 3-10 hrs.
Hardener 5-10%Starch to give wet resistance.
Disadvantages

1. Slow cure
2. Cationic resin coated - 10 - 14 days -full cure
   Anionic resin coated - 21 - 28 days -full cure
3. Hydrolyses readily at pH 4 - 6

Advantages

1. Cheap
2. Repulping possible
3. Retains property (wet strength) 6-8 years.

Use: Urea resin adhesive mixed with starch is used in making water resistant and high humidity resistant paper board boxes and corrugated paper board used in packing industries. Generally starch and dextrin pastes are used but these are not water resistant.Urea resin and hardener to the extent of 5 to 10% (dry wet) on the weight / starch / depending on the degree of water resistance desired and nature of the starch is used.

B. 2. **MELAMINE FORMALDEHYDE RESIN**

The melamine resins colloid are the first materials to be successfully applied to paper by introduction at the wet end of the paper making process.

The characteristics of a typical melamine colloid is as under:

Form-white, Free flowing Powder

Reactive resin content : 100%
Solid Content : 93%
Nitrogen Content : 38.9%
Shelf life : 12 months (See fig-1-3)

Melamine acid colloid solution is made as follows:
8.4 ml. of concentrated Hydrochloric Acid is added to 184-ml. of water heated to 35-38^{0}C. The acid solution is vigorously agitated; and 25 g of a spray-dried melamine formaldehyde powder resin (Molar ratio 1:3 MF) as above is added. When dissolution is complete the solution is allowed standing for 3 hours for the colloid to develop. The solid content of this colloidal solution is about 12%. It remains stable at least 3 days, but then rapidly increases in viscosity and gels. After dilution of the solution further to 6% solid, where the life will be indefinitely higher is to be added to a suitable point between the beaters and the head box. The quantity of resin will usually not exceed 3% solids on dry fibre.

Cationic Acid colloid solution has a direct affinity for cellulose pulp, which normally carries a negative charge, and the use of alum to assist retention is, therefore, unnecessary. The acidity of the stock may require adjustment however, and alum, hydrochloric acid or sodium carbonate may be used for this purpose according to circumstances. Optimum wet strength is obtained if the acidity of the stock is adjusted so that the pH of the paper as it leaves the machine is within the range of 4.5 - 5.0.
Between 70 and 85% of the resin added is retained in the paper web after formation and fairly rapidly on passing over the drying cylinders. A further improvement in wet strength usually occurs during storage if the paper is reeled hot.

Table 1 Comparison of U-F & M-F

Parameter	U-F	M-F
Rate of cure	Moderate	Fast
pH tolerance	Low	High
Nature of pulp	Selective	Any kind
Resistance of hydrolysis	Poor	Good
Beating period	More	Less
Effect of humidity	Dimensional change	No change
Sticking to Yankee drier	-	Low
Gluing problem	None	None
Resistance to delamination	Good	Excellent
Folding endurance	Good	Excellent
Dry strength	Increase	More increase
Stain resistance	Poor	Good
Abrasion resistance	-Poor	Good
Heat resistance	Poor	Good
Curing conditions	147 – 163°C in 3 min	130- 190° C in 1.5 min

Table 2 Application of Amino Resins By Beater Addition

	Bleached sulphite pulp				Un-bleached kraft pulp			
	Without alum		With alum		Without alum		With alum	
	Dry	Wet	Dry	Wet	Dry	Wet	Dry	Wet
Resin 3% weight	Tensile strength (1b/in)		Tensile strength (1b/in)		Tensile strength (1b/in)		Tensile strength (1b/in)	
None	5.4	1.2	5.3	1.2	11.0	1.2	10.2	1.2
U-F	6.8	1.7	9.0	3.3	15.8	3.6	20.5	8.2
Anionic U-F	10.0	5.3	10.0	5.3	21.4	7.5	21.0	9.1
Cationic M-F acid colloid	9.4	4.1	10.0	4.4	18.0	5.5	19.4	6.6

Applications of M.F. resins

M-F resins are used for making superior grade wet-strength paper wherever U-F coating is not suitable. One of the major applications of M.-F is to make decorative laminates. The various steps in making decorative laminates are given below:

1. Assemble 6-8 layers (core) of p-f bonded Kraft paper.
2. Put a M-F impregnated alpha-cellulose paper.
3. Put the decorative layer (0.008-0.010"): Highly refined alpha-cellulose Alpha-cellulose paper + 40-50% U-F or M-F resin.
4. Put protective overly (0.001-0.003"): pulp + 65-75% M-F resin.
5. Cure under heat and pressure.

Disadvantages of aminos

1. Acidic pH (around-4.5) during curing
2. Deterioration of paper:
   Embrittlement
   Reduction in Absorbency
   Ha-rsh Hand
3. Resins hydrolyze at low pH
4. Gradual wet-strength reduction during storage.

B.3 **POLYMINE AND POLYAMIDE WET- STRENGTH RESINS**

The amino-formaldehyde resins have the disadvantage of requiring acid conditions for curing. The recommended pH of around 4.5 causes embittlement and deterioration of paper, reduces absorbency and develops a harsh hand. These characteristics are especially undesirable in tissues and toweling. In addition, the wet-strength resins are hydrolyzed at low pH resulting in a gradual reduction of wet strength on storage, especially when the paper is wet. It was apparent that a wet-strength agent, which would cure at pH 7 or higher, was desired.

In 1952, Daniel and Landes patented a polymer of poly-alkylene Polyamine and epichlorohydrin as a wet-strength resin, which could be cured at pH levels as high as 10. Several years later, a polyamine-epichlorohydrin product was introduced to paper mills, and this type of resin has found general acceptance as a wet-strength agent for absorbent papers and other grades made at pH around 7. The resin is produced from an aliphatic dicarboxylic acid, such as, adipic acid, an alkylenepolyamins, such as, diethylene triamine, and epichlorohydrin. The product is an efficient wet-strength resin, which cures at pH higher than that required for the formal dehyde resins.

A typical polyamide resin produced from adipic acid, diethylene-triamine, and epichlorohydrin would have the structure:

```
                                   CH2-Cl
                                   1
                                   CH-OH
                                   1
                                   CH2
                                   1
                                                    O          O
                                    (+)             11         11
               ------HN - H2C - H2C - N - CH2 - CH2 - NH - C - (CH2)4 - C ------
                                   1
                                   CH-OH
                                   1
                                   CH2
                                   1
 O          O
 11         11                      (+)
-- C - (CH2)4 - C - HN - H2C - H2C - N - CH2 - CH2 - NH-------------
                                   1
                                   CH2
                                   1
                                   CH - OH
                                   1
                                   CH2 - Cl
```

Loss of hydrogen chloride is accelerated under basic conditions, and the resulting epoxide group is free to react with cellulose or other amino groups.

The polyamide resins, like other wet-strength resins, may be introduced to paper stock at any convenient location after the major refining work is done and before the sheet is formed, as for example, at the fan pump. The resins are retained efficiently, and absorption occurs rapidly. The resin may be added at any pH from 4 to 10, although the range 6-8 is usually preferred.

There are several significant differences between polyamide resins and the amino-formaldehyde resins. The polyamide resins cure over a broad pH range, and the rate of cure is affected by a lesser extent by change in pH. At the recommended pH conditions amino-formaldehyde resins reduce the absorbancy of paper, but the polyamide resins have little effect. At low levels of add-on, the polyamide resins are more efficient than the amino-formaldehyde resins, but higher ultimate wet strength can usually be obtained using urea-or melamine formaldehyde adducts. The polyamide resins are well retained without the use of aluminium sulfate although addition of 25-ppm sulfate ion is reported to help in achieving maximum efficiency at high levels of addition.

Paper containing U-F and M-F wet strength agents is usually repulped with acid, whereas that containing polyamide resins require hydrochlorite or a high pH. The amino resin increases the dry strength more than do the polyamide resin.

Amino polyamide-epichlorohydrin resins are significantly more expensive than formaldehyde based resins, but the higher cost is offset by other advantages, mostly the advantage of making paper at neutral to alkaline pH. These include properties, such as, improved softness and absorbency, no tendency towards embrittlement on aging, and reduced paper machine corrosion. Another important advantage is the elimination of formaldehyde usage.

Characteristics

1.	Gel time (200 g)	:	2-4 hr.
	Viscosity	:	120 Ns /m2 at 210°C
	Shelf life	:	6 months, 200°C
	Curing pH	:	Higher than Aminos

Table 3 Comparison With Aminos

Parameters	Amino	PA
ph range	Acidic	4 - 10
Rate of cure (effect of pH)	Great	No significant
Paper Absorbency	Reduction	No effect
Wet- strength (at comparable add on)	High	Higher
Alum	Necessary	Not necessary
Repulping conditions	Normal	Special
Dry strength	Increase	More increase
Softness	Less	More
Aging	Brittle	No change
Machine corrosion	High	Low
Cost	Low	High

B-4 GLYOXAL RESINS

Another class of wet strength resin is based on the reaction product of polyacrylamide and glyoxal. This resin can react with hydroxyl groups on cellulose fibers through the glyoxal functionality to form hemiacetal bonds, resulting in improvements in paper wet strength, although, not to the same extent as amino-polyamide-epichlorohydrin resins. The base polyacrylamide usually contain a small amount of cationic co-monomer, such as, dimethyldiallyl ammonium chloride to make the resin substantive to pulp fibers.

```
------- CH2  -   CH   -  CH2  -  CH    -  CH2  -   CH ----------------
                 1               1                  1
                 C=0             C=0                C=0
                 1               1                  1
                 NH2             NH                 NH2
                                 1
                                 CH2
                                 1
                                 OH
```

B - 5 DIALDEHYDE STARCH

Dialdehyde starch, made by periodic oxidation of cornstarch, is the newest type of resin to achieve commercial attention as a wet-strength resin. Dialdehyde starch is anionic in nature and can be used as wet-end additive only if the paper fibers have been treated with cationic starch or synthetic cationic retention acids. High amounts of aluminum sulphate are required for retention of dialdehyde starch. Dialdehyde starch produces only a temporary wet strength, but cationic dialdehyde starch confers more permanent wet strength. Temporary wet strength is useful in towels and sanitary tissues where wet strength is required for only a few seconds and where Permanent strength can cause clogging of plumbing and septic tanks. In addition to temporary wet strength, dialdehyde starch offers extremely easy broke recovery, high wet strength without loss in absorptivity, rapid development of wet -strength, and excellent improvement in dry strength properties. It is the only paper additive that imparts significant improvement in wet and dry strength without causing broke recovery problem.

B - 6 **CHITOSAN**

Chitosan (poly 2- aminodeoxy - 1,4 glucoside) is not a single substance, but a group of partially deacetylated products of various degrees of polymerization. Chitosan is insoluble in water, concentrated alkalies, alcohol, and acetone, but dissolves readily in dilute acid. It has also been shown to impart wet strength to paper. The polymer, although potentially available in large quantities from shell-fish wastes has never become a commercial significant product.

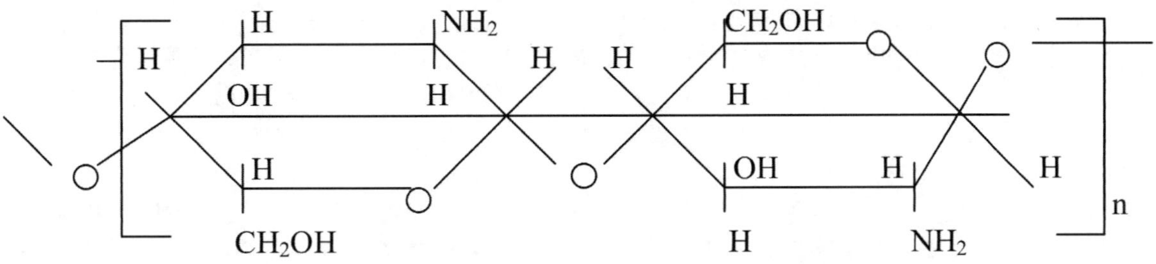

B -7 **MISCELLANEOUS**

In addition to the urea-formaldehyde, melamine- formaldehyde, polyamide, and dialdehyde-starch resins, other resins may impart some wet strength, but are used primarily for purposes other than wet strength.

Small amount of neoprene latex added at the wet end of the papermaking machine gives high wet strength and resistance to a variety of chemicals. Saturation of paper with styrene- butadiene, acrylic or butadiene arylonitrile latex for use in manufacturing book covers, marking tape and gaskets confers wet strength and other properties to paper. Latexes also change many other properties of paper, e.g., hand (the feel of the paper), elongation, stiffness, dimensional stability, edge and internal tearing resistance, folding endurance and tensile strength. Therefore, they are not used when wet strength improvement is the only change desired.

Because latex particles are not readily retained by pulp; their use is frequently more troublesome, but aluminium sulphate aids their retention. Saturation with phenolic resin, like saturation with latex, produces a resin-filled sheet that has high wet strength. However, the treatment of paper with latex or phenolic resin changes its properties in such a fundamental way that it is not usually called " wet- strength treatment".

B- **8 NEWER RESINS**

Polyethylenimine was used in Germany during World War-II as a wet strength agent. It has never become commercially significant, however, it confers several unique properties. It retards the rate of beating of pulp and should be added after the pulp is beaten. Wet strength develops as the pulp is dried, and water absorbency is not affected.

Cereal xanthates, produced from cereal starches or flours, produce permanent wet strength and increase dry tensile strength and folding resistance.

Polyacrolein gives paper with high, permanent wet strength and good dimensional stability. The polycryolein must be treated with sulfur dioxide or a sulphite to make it water-soluble.

C. MECHANISM OF WET STRENGTH FORMATION

With the exception of polymeric amines, all effective wet- strength resins undergo cross-linking reactions and some react with functionality present on the surface of cellulose pulp fibers. Two mechanisms for wet strength development have been proposed.

The protection theory suggests that the resin forms a cross-linked network around existing fiber-fiber bonds and prevents swelling and disruption of these bonds.

The new bond theory proposes that the wet strength resin creates new fiber-fiber bonds that are covalent and thus are not broken on exposure to water. The protection theory is supported by the fact that a given concentration of amino polyamide- epichlorohydrin resin imparts wet strength that is a relatively constant percentage of dry strength in various pulp furnishers, even though the absolute values of wet and dry strength might vary substantially.

The new bond- theory is supported by the fact that wet and dry strength are increased by approximately the same increment with either formaldehyde- based or amino polyamide-epichlorohydrin wet- strength resins.

D. REPULPING

Repulping of paper is necessary for two reasons. During paper manufacture broke (waste) is produced either during breaks on the paper machine or as trim produced during the rewinding process. For the paper making process to be economical, this must be recycled. Also, the use of waste paper furnishes is increasing as the cost of virgin-pulp furnishes increase.

Paper containing wet- strength resin is more difficult to repulp than non-wet-strength paper. Therefore, the repulpability properties of wet strength resins must be considered when choosing a resin. U- F and M-F resins can be readily repulped under acid conditions. Alum is used in the repulping operation when these resins are used. Paper containing polyamide- epichlorohydrin resins are not readily repulped under either acidic or basic condition. Oxidizing agents, such as, sodium hypochlorite is usually required for repulping in such cases. Presumably, the oxidant attacks the polyamide backbone and breaks down the resin. Paper containing poly (methyl diallyamine) epichlorohydrin wet-strength resin is very difficult to repulp. The hydrocarbon backbone of this resin is not subject to attack by reagents commonly used in aqueous systems.

E .APPLICATION OF RESIN COATING IN COLOUR ELECTROPHOTOGRAPHY

A. Basic idea

A recent suggestion has been made that emphasis in the dye and pigment industry is now shifting from the replacement of natural products, quality control and process improvements. Some newer ideas, such as, colour- formers for carbonless copy paper are already in the market place and dyes are becoming of increasing use in information technology.

Recording and reproduction of information using photography and photocopying are already well known. Laser- writing and reading, liquid- crystal displays and the use of compact discs, in all these processes, at some stage either in manufacture or use, electromagnetic radiation is absorbed and hence, can be influenced by the presence of dye or pigment. Other applications, which may prove more beneficial to society in the long run, may be found in converting solar to electrical energy and in dye- related drugs.

B. **Principle**

The principle involves in carbonless copy processes and colour formers is the rupture of microcapsules by impact and release of a virtually colorless pre-cursor which can interest with an underlying copy paper layer to form a colour.

Microcapsules are very small (1-500 μm dia) containers in which a fluid or solid is enclosed in a polymer shell. Spherical microcapsules containing liquid are more easily ruptured than those containing solid. They have many useful functions and are to be found in many different fields. The reasons for micro encapsulation include the following:

a) Protection of the enclosed material and improved storage life.
b) Conversion of a liquid component to a dry solid system.
c) Ensuring separation of incompatible components.
d) Odour masking, dust control and pH control, and,
e) Controlled diffusion of active component through the shell, as for example, in delayed drug release. The general principle of micro-encapsulation is shown in Fig.-4

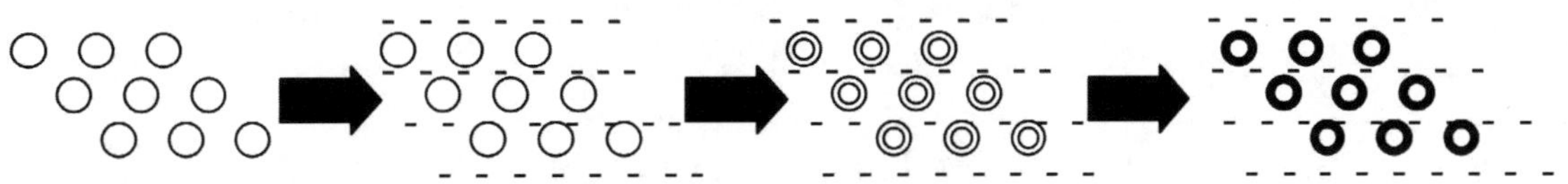

a) Dispersion or emulsion | b) Wall material dissolved in continuous phase | c) Wall material deposited on core | d) Deposited wall solidified & hardened

Formation of Microcapsules.

Phase separation results from the interaction of two oppositely charged, water-soluble polymers, such as, gelatin and gum Arabic. Both polymers exist as negatively charged species in neutral aqueous solution. Where this mixed solution forms the continuous phase in an oil - in - water emulsion and when the pH is reduced to 4.8 only the gelatin becomes positively charged and mutual electrostatic attraction between the polymers causes precipitation around the organic droplets.

For microcapsule use in carbonless copy paper the organic phase, which may be chlorinated paraffin contains a colorless dry pre-cursor known as a colour former. Other means, such as, interfacial polymerization to form the polymer shell, in which the reacting monomers are separately contained in the oil and water phases of the emulsion, is a later development.

Among organic coating are to be found phenol-formaldehyde resins, bisphenols, such as, 2, bis (p-hydroxyphenyl) propane and 4-alkoxy - or 4-phenoxy-1- naphthol. Phenolic resins in which some of the phenolic positions have been replaced by zinc give improved shelf life and inhibit yellowing.

C. **APPLICATION OF POLYMER COATED PAPERS IN ELECTROPHOTOGRAPHY.**

1. Colour photocopying and photography:

A recent development is colour photocopying which does not employ electrostatic attraction of toners. The method makes use of microcapsules containing, in addition to a colour former, a monomer and photosensitive initiator. By the suitable choice of initiator the capsules can be made sensitive to blue (420 mm), green (550 mm) or red (650 mm) radiation. After exposure, the cores of those capsules, which have been polymerized, do not release their colour formers when subject to pressure. By coating a paper with all three types of capsule and exposing under a transparent document, or under a transmissive liquid crystal display, a latent colour image is produced. This image is then placed in face-to face contact with an acidified clay- coated layer and squeezed through the nip of pressure rollers to produce a final colour print on the clay coated-layer.

REFERENCES

1. Mark, H.F., Goylord, N.G., Bikales, N.M., "Encyclopedia of Polymer Science and Technology." Interscience Publishers, NY, Vol. 9, 748-93 (1965).

2. Mark, H.F., Mcketta, J.J., Othemer, D.F., "Encyclopedia of Chemical Technology." Interscience Publishers, NY, Vol. 14, 500-31 (1963).

3. Mark, H.F., Gaylord, N.G., Bikales, N.M., "Encyclopedia of Polymer Science and Technology." Interscience Publishers, NY, Vol. 12, 494- 531 (1987).

4. Shields, J., " Adhesives Handbook", Published by Butter Worths, London, 74-5 (1970).

5. Vale, G.P., Taylor, W.G.K., "Amino plastics," Published by ILIFFE Book Ltd., London, 113-20 (1964).

6. Jones, F., Rev. Prog. Coloration, Vol. 19, 20 - 32 (1989)

7. Kokurek M.J., Pulp and paper Manufacture, Vol.7 , Joint text book Committee of the Paper Industry, TAPPI Press, 1989

8. Biermann, C.J., 1996, "Handbook of Pulping and Papermaking", Academic Press, CA.

9. Scott W.E., Principles of Wet End Chemistry, TAPPI Press, 1996

10. Personal Communication with Prof. A.K.Mukherjee, IIT Delhi,1995

Session 17006 Recalcitrance of Woody Biomass
AIChE 2012
October 28 ~ November 2
Pittsburgh, PA

A study in lignin inhibition for enzymatic hydrolysis of woody biomass

Zhiying Yu*, Hasan Jameel, Hou-min Chang, Sunkyu Park

Department of Forest Biomaterials, North Carolina State University
Campus Box 8005, Raleigh, NC 27695-8005, USA
Presenting author: zyu2@ncsu.edu

The presence of lignin in woody biomass is a great obstacle that limits the bioconversion efficiency through enzymatic hydrolysis, especially for softwood. It has been speculated that the differences in lignin content, structure, and distribution between hardwood and softwood, will influence lignin-enzyme interactions such as non-specific adsorption of enzyme onto lignin, physical blockage of lignin that restrict the enzyme penetration into biomass structure, and chemical blockage of binding sites through lignin-carbohydrate linkages. In this research, the milled wood lignins (MWL) isolated from eucalyptus, maple and loblolly pine as well as kraft lignin (indulin AT) was investigated for their impact on enzymatic hydrolysis. The woody biomass structure was mimicked by 1) physically mixing the lignins and bleached pulps and, 2) reconstructing the woody biomass via impregnating the lignins into bleached pulps.

The adsorption kinetics of cellulase on individual bleached pulps and lignins showed that the bleached pulps adsorbed 2 to 3 folds of cellulase than lignins. MWL adsorbed 17.5-22.5% cellulase, while kraft lignin only had less than 3.5 % cellulase adsorption. When 20% lignins were physically mixed with bleached pulps for enzymatic hydrolysis, it was found that the presence of 20% MWL had a minor impact on enzymatic hydrolysis, while 20% kraft lignin showed 8.3-11.6% limitation on the hydrolysis. In contrast, when 20% lignins were impregnated into the bleached pulps, 6.2%, 12.0%, 17.6% and 59.0% decrease in 96 h hydrolysis yield was observed for reconstructed eucalyptus, maple, pine and kraft pulps, respectively. Therefore, MWL isolated from untreated hardwood and softwood was found to limit enzymatic hydrolysis due to the effect of physical blockage. A more significant impact of physical blockage of lignin was found with the increasing of guaiacyl unit of lignins in the reconstructed woody biomass with the same lignin content. Kraft lignin inhibited the enzymatic hydrolysis through slightly deactivating the enzymes and significantly blocking the accessible area.

It was speculated that the smaller molecules of lignin had a greater physical blockage effect on enzymatic hydrolysis. Thus, different molecular weight of kraft lignins was fractionated by selective pH and ultrafiltration, respectively. The lignins of bigger molecular weight were found to have less physical blockage than the lignin with smaller molecular weight. In order to investigate the effect of lignin from pretreated lignocelluloses on enzymatic hydrolysis, the cellulolytic enzyme lignins were isolated from acid pretreated and alkali pretreated hardwood and softwood pulps. The effect of acid and alkali pretreatment on lignin structure as well as the effect of hardwood and softwood lignins on enzymatic digestibility will be discussed in terms of three inhibitory mechanisms suggested above.

(787b) Distinct Effects of Lignin and Xylan On the Enzymatic Hydrolysis of Organosolv Pretreated Biomass

Mi Li, School of forestry and wildlife science, Auburn University, Auburn, AL and Maobing Tu, Auburn University, Auburn, AL

Forestry lignocellulosic biomass is a great potential source of sustainable materials for biofuels and chemicals. Technically, bioconversion process of cellulosic ethanol from forestry feedstock requires pretreatment, enzymatic hydrolysis and fermentation. In order to improve accessibility of lignocellulose to enzymes for hydrolysis, high temperature pretreatment with acid, base or solvent is typically needed to break down the recalcitrant structure of cellulose. Besides the degree of cellulose crystallinity, the interaction between xylan/lignin and cellulase enzymes played a key role in the effective hydrolysis of lignocellulosic biomass. Elucidation of the distinct roles of residual xylan and lignin has been investigated in this study. Loblolly pine (*Pinus taeda*) and sweetgum (*Liquidambar*) were pretreated by organosolv and have been quantitatively characterized by the enzymatic hydrolysis based on initial hydrolysis rate and final hydrolysis yield. By correlating the residual lignin and xylan to the initial hydrolysis rate and final yield in organosolv pretreated softwood and hardwood, a more accurate fundamental understanding the roles of xylan and lignin in limiting the enzymatic hydrolysis has been developed. Our results show that Novozyme 22C showed very good performance of hydrolysis on organosolv pretreated biomass. Moreover, the residual xylan in pretreated substrates actually would affect the initial hydrolysis rate much more than that from residual lignin. It suggested that xylan firstly inhibited the cellulase because it's closer to the cellulose micro-fibril. The lignin residual interacted with cellulase after xylan and decreased the overall hydrolysis yield. The addition of xylanases could increase the initial hydrolysis rate by removing xylan. We also found that in the simultaneous saccharification and fermentation the ethyl xylopyranoside was produced by the enzymatic catalysis of xylose and ethanol.

(787c) Ionic Liquid Pretreatment of Lignocellulosic Biomass – Recovery and Reuse of IL

Anantharam P. Dadi, SuGanit Systems Inc, Toledo, OH, Marcus Stacy, SuGanit Systems Inc., Toledo, OH, Guneet Kumar, SuGanit Systems Inc., Reston, VA, Noureen Faizee, Synterra, Toledo, OH and Sasidhar Varanasi, Chemical & Environmental Engineering, University of Toledo, Toledo, OH

University of Toledo and SuGanit Systems Inc. have jointly developed a novel Ionic Liquid pretreatment method for deconstructing lignocellulosic biomass feedstocks to promote rapid enzyme-catalyzed saccharification of the carbohydrates . The IL pretreatment is a facile method that works at moderate temperatures and atmospheric pressure conditions, and was successfully demonstrated on various lignocellulosic biomass feedstocks. SuGanit Systems Inc. is scaling up this technology of sugar production from lignocellulosic biomass by ionic liquid pretreatment. A critical economic driver of the process is efficient recovery of IL from wash streams following biomass pretreatment. We have explored the viability of electrodialysis (ED) for recovering and concentrating IL following biomass pretreatment. It was shown that dissolved nonionic extractibles entering IL wash solutions from biomass could successfully be removed via ED. This presentation will focus on the effectiveness and economics aspects of ED for this purpose, and the important differences between ED-based separation of IL versus salt solutions.

(787d) Solvent Based Pretreatment (N-methyl morpholine oxide) for Enhanced Enzymatic Hydrolysis of Dissolving Pulp and Sugarcane Bagasse

Subramanian Ramakrishnan, Department of Chemical and Biomedical Engineering, Florida A&M University - Florida State University, Tallahassee, FL, Gary Brodeur, Chemical & Biomedical Engineering, Florida State University, Tallahassee, FL, John Telotte, Department of Chemical and Biomedical Engineering, Florida State University, Tallahassee, FL and John Collier, Chemical and Biomedical Engineering, Florida State University, Tallahassee, FL

The first step in the conversion process of biomass to biofuels or hydrogen lies in picking an environmentally friendly solvent which is capable of breaking up the crystalline microstructure of cellulose. Once the cellulose is in suspension, enzymes (cellulases) are added to the reactor to hydrolyze the cellulose to simple sugars that are then fermented by bacteria to biofuels. Till date, a number of solvents have been used for pretreatment of biomass and for reprecipitating it by the addition of an anti-solvent thus making it a multi-stage process. In this work, it is shown that N-methyl morpholine oxide is a good solvent for pretreatment of crystalling pulp and sugarcane bagasse. Sugarcane bagasse and dissolving pulp of degree of polymerization 1160 are used as the substrate and Accellerase 1000 and 1500 purchased from Genentech are the cellulases used. The rates of enzymatic hydrolysis and total sugars released are studied as a function of solution conditions – solution pH, temperature and enzyme loading. These studies reveal that the enzymes are active in NMMO and that the rates of hydrolysis of cellulose pretreated with NMMO are comparable to the rates of hydrolysis of regenerated cellulose suspended in aqueous buffer solutions. A comparison of enzymatic hydrolysis after NMMO treatment of sugarcane bagasse reveals that the rates and yields are comparable and in certain instances higher than acid pretreatment followed by steam explosion. A mathematical model to predict enzymatic hydrolysis based on lumped kinetic modeling is also proposed.

(787e) Role of Tween-80 in Reduction of Nonproductive Cellulase Binding to Lignin

Ravinder Garlapalli[1], Barbara, L Knutson[2], Sue E. Nokes[3] and Stephen E. Rankin[1], (1)Chemical and Materials Engineering, University of Kentucky, Lexington, KY, (2)Chemical & Materials Engineering Department, University of Kentucky, Lexington, KY, (3)Department of Biosystems and Agricultural Engineering, University of Kentucky, Lexington, KY

A key limitation in the enzymatic hydrolysis of lignocellulose to produce soluble sugars is the nonproductive binding of the enzymes to biomass components such as lignin. Bulk studies on hydrolysis of cellulosic substrates have suggested that additives enhance both conversion of sugars and enzyme recovery when lignin is present, where nonionic surfactants are the most widely studied additives. However, the specific mechanism underlying the use of surfactants is not well understood. To study the effect of nonionic surfactant Tween-80 and the interactions with cellulase enzyme on lignin, model lignin thin films are prepared and investigated as a function of the adsorbing solution using a Quartz Crystal Microbalance with Dissipation monitoring (QCM-D). Model lignin films were synthesized by dissolution of dry lignin powder in ammonium hydroxide solution followed by coating onto a quartz crystal resonator. The cellulase for this study is a commercially available mixture of enzymes derived from *Trichoderma reesei*. In-situ studies of single-component binding show that cellulase is bound to the lignin surface irreversibly, while Tween-80 is partially (but not completely) reversible upon rinsing with buffer. Co-adsorption of Tween-80 and cellulase (where both are introduced together) resulted in an apparent net reduction in the amount of cellulase adsorbed on lignin. Sequential adsorption experiments were carried out by varying the order and concentration of Tween-80 and cellulase introduction to lignin surfaces to explore the possibility for displacement of cellulase or pretreatment of biomass using Tween-80. The results of the sequential adsorption experiments suggest that Tween-80 was able to displace adsorbed cellulases, but quantifying the amount of cellulase displaced was difficult with QCM-D as it measures the net change in piezoelectric properties due to mass adsorbed and desorbed on the sensor. Further studies are underway to use analytical methods to quantify the amounts of cellulase desorbed under various conditions and to better understand the mechanism by which the surfactants reduce the nonproductive binding of cellulases through their interaction with lignin.

(787f) Green and High Efficient Fuel Ethanol Production From Sugarcane Bagasse

Xinshu Zhuang, Bio-chem conversion lab, Guangzhou institute of energy conversion,Chinese Academy of Sciences, Guangzhou, WI, China

Sugarcane bagasse (SCB), leftover stalks after juice extraction, were pretreated using liquid hot water (LHW) to remove most hemicellulose and part lignin of lignocellulose. In order to improve the enzymatic hydrolysis efficiency of LHW-pretreated SCB, Tween80, a non-ionic surfactant proved to enhance enzymatic hydrolysis of pretreated biomass in high lignin content, was applied in this research. It was found that the optimal concentration of Tween80 can increase glycan conversion of LHW-pretreated SCB from 44.9% (control) to 78.1%. A laboratory facility with plate-and-frame impeller (PFI) was applied to conduct batch and fed-batch hydrolysis and the process of fed-batch hydrolysis prior to simultaneous saccharification and fermentation (SSF). After hydrolyzing LHW-pretreated SCB for 120 h, the total sugar concentration and glycan conversion obtained from fed-batch hydrolysis are 91.6 g/L and 68.3%, respectively, which are 9.7 g/L and 7.3% more than those from batch hydrolysis. The ethanol production and theoretical yield obtained from the process of fed-batch hydrolysis prior to SSF are 55.4 g/L and 88.3%, respectively, which are 15.5 g/L and 24.7% higher than those from separate fed-batch hydrolys and fermentation.

These research work are financially supported by the National Key Technology R&D Program (2011BAD22B01).

(787g) Crystallinity – A Rate-Determining Parameter for Pretreated Lignocellulosic Biomass

Yuzhi Kang[1], Prabuddha Bansal[1], Bryan Vowell[2], Matthew Realff[1], Jay H. Lee[3] and Andreas S. Bommarius[4], (1)School of Chemical and Biomolecular Engineering, Georgia Institute of Technology, Atlanta, GA, (2)Georgia Institute of Technology, Atlanta, GA, (3)Chemical and Biomolecular Engineering, Korea Advanced Institute of Science and Technology, Daejeon, South Korea, (4)Chemical & Biomolecular Engineering, Georgia Institute of Technology, Atlanta, GA

Crystallinity – A Rate-determining Parameter for Pretreated Lignocellulosic Biomass

Yuzhi Kang, Prabuddha Bansal, Bryan Vowell, Matthew Realff, Jay Lee, Andreas Bommarius

Abstract

Steam explosion is one of the most effective pretreatment methods in disrupting lignocellulosics structure and enhancing its accessibility. The effects of steam explosion on woody biomass and agricultural residues (i.e. bagasse, wheat straw and loblolly pine) are evaluated. The key operational parameters including temperature, residence time and acid concentration have pronounced effects on crystallinity index (CrI) of the feedstock. The CrI of steam-pretreated lignocellulosic biomass was calculated from X-ray diffractometry with the method of Least-squares and was found to decrease as pretreatment severity increases. A series of steam-exploded lignocellulosic samples with varying CrI was generated and the significance of CrI is confirmed by a linear relationship between initial hydrolysis rates and CrI. The present study provides evidence that the CrI is one of key parameters determining the enzymatic hydrolysis rate of pretreated lignocellulosic biomass.

To more fully explore the connection between accessibility, hydrolyzability and reactivity, we have undertaken experiments with phosphoric acid swollen Avicel® at different levels of conversion. We will present the quantitative trends between the rate of conversion and these underlying substrate properties.

AUTHOR INDEX

AUTHOR INDEX